Key Factors in Postgraduate Research Supervision
A Guide for Supervisors

RIVER PUBLISHERS SERIES IN INNOVATION AND CHANGE IN EDUCATION - CROSS-CULTURAL PERSPECTIVE

Volume 11

Series Editor

XIANGYUN DU
Aalborg University
Denmark

Editorial Board

- **Alex Stojcevski**, Faculty of Engineering, Deakin University, Australia
- **Baocun Liu**, Faculty of Education, Beijing Normal University, China
- **Baozhi Sun**, North China Medical Education and Development Center, China
- **BinglinZhong**, Chinese Association of Education Research, China
- **Bo Qu**, Research center for Medical Education, China Medical Education, China
- **Danping Wang**, The Department of General Education, Technological and Higher Education Institute of Hong Kong
- **Fred Dervin**, Department of Teacher Education, Helsinki University, Finland
- **Kai Yu**, Faculty of Education, Beijing Normal University, China
- **Jiannong Shi**, Institute of Psychology, China Academy of Sciences, China
- **Juny Montoya**, Faculty of Education, University of ANDES, Colombia
- **Mads Jakob Kirkebæk**, Department of Learning and Philosophy, Aalborg University, Denmark
- **Tomas Benz**, Hochschule Heilbronn, Germany

Nowadays, educational institutions are being challenged when professional competences and expertise become progressively more complex. This is mainly because problems are more technology-bounded, unstable and ill-defined with the involvement of various integrated issues. To solve these problems, it requires interdisciplinary knowledge, collaboration skills, innovative thinking among other competences. In order to facilitate students with the competences expected in professions, educational institutions worldwide are implementing innovations and changes in many aspects.

This book series includes a list of research projects that document innovation and change in education. The topics range from organizational change, curriculum design and innovation, pedagogy development, to the role of teaching staff in the change process, students' performance in the aspects of not only academic scores, but also learning processes and skills development such as problem solving creativity, communication, and quality issues, among others. An inter- or cross-cultural perspective is studied in this book series that includes three layers. First, research contexts in these books include different countries/regions with various educational traditions, systems and societal backgrounds in a global context. Second, the impact of professional and institutional cultures such as language, engineering, medicine and health, and teachers' education are also taken into consideration in these research projects. Thirdly, individual beliefs, perceptions, identity development and skills development in the learning processes, and inter-personal interaction and communication within the cultural contexts in the first two layers.

We strongly encourage you as an expert within this field to contribute with your research and make an international awareness of this scientific subject.

For a list of other books in this series, www.riverpublishers.com

Key Factors in Postgraduate Research Supervision
A Guide for Supervisors

Dario Toncich (PhD), (MEng), BEng(Elec)(Hons)

LONDON AND NEW YORK

Published 2016 by River Publishers
River Publishers
Alsbjergvej 10, 9260 Gistrup, Denmark
www.riverpublishers.com

Distributed exclusively by Routledge
4 Park Square, Milton Park, Abingdon, Oxon OX14 4RN
605 Third Avenue, New York, NY 10017, USA

Key Factors in Postgraduate Research Supervision: A Guide for Supervisors / by Dario Toncich.

© 2016 River Publishers. All rights reserved. No part of this publication may be reproduced, stored in a retrieval systems, or transmitted in any form or by any means, mechanical, photocopying, recording or otherwise, without prior written permission of the publishers.

Routledge is an imprint of the Taylor & Francis Group, an informa business

ISBN 978-87-93379-44-2 (print)

While every effort is made to provide dependable information, the publisher, authors, and editors cannot be held responsible for any errors or omissions.

"Nothing in life is to be feared, it is only to be understood. Now is the time to understand more – so that we may fear less."

– Marie Curie

Contents

Foreword   xv

Acknowledgments   xix

List of Figures   xxi

List of Tables   xxiii

List of Abbreviations   xxv

PART I: Preparation

1 Introduction and Text Usage Guide   3
    1.1 Aims and Text Structure 3
    1.2 Text Overview 9
    1.3 Overview of the Research Supervisory Process 12

**2 Understanding the Fundamental Outcomes
of Research Supervision**   15
    2.1 Overview 15
    2.2 Development of Fundamental Research Skills 18
    2.3 Execution of a Structured Program of Research 20
    2.4 A Contribution to (Extension of) Knowledge
        in a Particular Field 23
    2.5 Development of Rigorous Documentation Techniques
        and Disciplined Writing Style 25
        2.5.1 Introduction 25
        2.5.2 Starting with Basic Research Objectives – A Central
              Research Theme 28

		2.5.3	Developing Writing Discipline Early in the Research Program	31
		2.5.4	The Issue of Grammar and Writing Competence . .	34
	2.6	Development of Oral/Visual Research Presentation Skills		37
	2.7	Peer Evaluation/Acceptance of Research		39
	2.8	Published Research Findings		41
	2.9	An Understanding of the Broader Perspective of the Research and Research Methods		43
	2.10	Establishment of a High Caliber Research Team in Which the Research Student Is an Active Member		44
	2.11	Creation and Maintenance of a Dynamic Research and Learning Environment Which Actively Encourages the Exchange of Ideas and Opinions		45
	2.12	Establishment of a Research Team in Which Peer Review of Each Member's Work Is an Intrinsic Part of the Research Effort		46
	2.13	A Dissertation/Thesis for Examination and Defense		47
		2.13.1	Nomenclature	47
		2.13.2	The Supervisor's Role in Thesis Preparation	47
	2.14	A Professional Career Foundation		49
	2.15	Underpinning Research Outcomes for Future Research Grants		51
	2.16	Research Outputs that Feed into University Research Performance Metrics		52
	2.17	Establishment of a Long-Term Professional Relationship between the Graduated Student and the Supervisor		53
3	**Basic Responsibilities of Research Supervision**			**55**
	3.1	Overview		55
	3.2	Health/Safety		56
	3.3	Personal Welfare of the Research Student		60
	3.4	Confidentiality of the Research Student and Research		64
	3.5	Protection of Intellectual Property (IP)		65
	3.6	Research Ethics		67
	3.7	Prevention of Discrimination, Harassment/Bullying		69
	3.8	Personal/Professional Ethics Development		72
	3.9	Personal Development		73
	3.10	Selection of Research Examiners		76

4 The Relationship between the Supervisor and the Student — 79
4.1 The Role of the Supervisor 79
4.2 Supervisor Types 81
4.3 Interaction between Students and Supervisors 84
4.4 Typical Supervisor/Student Problems 87
4.5 Conflict Management and Resolution of Disputes 89

5 Understanding the Research Environment — 95
5.1 Overview 95
5.2 A Perspective on Global University Numbers 96
5.3 International Rankings 97
    5.3.1 General Issues 97
    5.3.2 How International Rankings Affect Research Supervisors 100
5.4 Institutional Funding Arrangements 102
5.5 Institutional Structures 105
5.6 University Governance Structures 106
5.7 Internal Performance Metrics 109

PART II: The Supervisory Process

6 Preliminary Tasks in Research Supervision — 113
6.1 Overview 113
6.2 Definition/Funding of Research Projects 113
6.3 Scholarship Funding 116
    6.3.1 General Issues 116
    6.3.2 Scholarships Arising from Collaborative Research Programs 118
    6.3.3 Other Issues 119
6.4 Student Recruitment 120
    6.4.1 General Issues 120
    6.4.2 Selecting the Appropriate Candidate for Postgraduate Research 121
6.5 Submission of Candidature Forms 124
    6.5.1 General 124
    6.5.2 Candidature Processes 125
    6.5.3 Ethics Considerations 127
    6.5.4 Timeline Considerations 127

x Contents

7 Initiating the Formal Supervisory Process — 129
- 7.1 Introduction — 129
- 7.2 General Induction for Research Students — 132
- 7.3 Occupational Health/Safety — 132
- 7.4 Initial Meeting with the Research Student — 134
  - 7.4.1 General — 134
  - 7.4.2 Understanding the Research Student's Perspective — 135
  - 7.4.3 Negotiating a Mutually Acceptable Supervisory Approach — 136
  - 7.4.4 Establishment of Formal and Informal Meeting Mechanisms — 139
  - 7.4.5 Providing Constructive Feedback to Research Students — 141

8 Planning the Postgraduate Research Program — 145
- 8.1 Introduction — 145
- 8.2 Understanding a Deadline — 146
- 8.3 Basic Elements of a Postgraduate Research Program — 148
- 8.4 The Planning Process — 150
- 8.5 Using the Planning Process — 153

9 Conflict Resolution in Postgraduate Research — 155
- 9.1 Introduction — 155
- 9.2 Jurisdiction and Scope of Conflicts — 159
- 9.3 Not All Conflicts Can or Should Be Resolved — 160
  - 9.3.1 Basic Issues — 160
  - 9.3.2 Managing Disputes Relating to Misfeasance/Wrongdoing from a Senior Staff Member — 162
  - 9.3.3 Resignation from the University — 164
  - 9.3.4 Managing Disputes Relating to Misfeasance/Wrongdoing from a Research Student — 166
- 9.4 Conflicts with Committees — 169
- 9.5 Conflicts Involving Departmental/Faculty Resources and Technical Support — 171
- 9.6 Conflicts with External Partners — 173

10 Research Supervision in Industry/Partnered Collaborative Research — 177
- 10.1 Overview — 177
- 10.2 Contracts and Distribution of Liability — 179

10.3		Joint Management/Supervision	182
	10.3.1	General	182
	10.3.2	Project Management Committee	182
	10.3.3	Supervision of the Research Candidate	183
10.4		Definition of Outcomes/Deliverables	185
10.5		Utilization/Management of Resources and Funding	188
	10.5.1	Fundamentals	188
	10.5.2	Accountability/Auditing	189
	10.5.3	Resources	191
10.6		Preparation of Research Student for Collaborative Research	194
	10.6.1	Overview	194
	10.6.2	Basic Induction into the Company – Health and Safety	195
	10.6.3	Dress Codes	196
	10.6.4	Dealing with Company Staff	196
	10.6.5	Organizing Meetings with Company Staff	197
	10.6.6	Confidentiality	198
	10.6.7	Intellectual Property (IP)	198
	10.6.8	Bullying and Harassment	198
10.7		Health and Safety	199
10.8		Adherence to Multiple (Potentially Conflicting) Instructions and Procedures	200
10.9		Confidentiality	204
10.10		Intellectual Property (IP)	207
10.11		Student Welfare	210

11 Peer Review by Publication   213

11.1	Overview	213
11.2	When to Publish	214
11.3	How Often to Publish	215
11.4	Where to Publish	216
	11.4.1 General	216
	11.4.2 Online Preprint Servers	218
11.5	What to Do If Publication Is Not Practical	219

12 Preparation of a Thesis   221

12.1	Overview	221
12.2	Understanding the Thesis Readers/Examiners	222
12.3	Thesis Structure – Developing a Thesis Template	224

12.4 Flow of Argument Complexity 226
12.5 The Central Research Theme 229
12.6 Thesis Preparation Timeline 232
12.7 Writing Ability/Grammar 235
    12.7.1 General 235
    12.7.2 English Language Support Units 237
12.8 Documenting the Literature Review 238
    12.8.1 General 238
    12.8.2 Limitations of the Literature Review 241
    12.8.3 The Funneling Process 243
    12.8.4 Identifying Landmark Research
           and Seminal Authors 244
    12.8.5 The Literature Review as an Impetus
           for Research 245
    12.8.6 Writing the Review Chapter 245
12.9 Balancing a Thesis 246
    12.9.1 The Thesis Body 246
    12.9.2 Thesis Appendices 249
12.10 Personal Opinions, Lazy Phrasing/Numerical Phrasing ... 250
12.11 Creating a Cohesive Document 253
12.12 Understanding the Broader Context of the Research 256
12.13 Understanding the Concluding Chapter 257
12.14 Understanding the Abstract 259
12.15 Summarizing the Research Supervisor's Role
     in Thesis Preparation 261

PART III: Relevant Supervisory Issues

13 Research Misfeasance Issues   265
13.1 Overview 265
13.2 Misrepresentation of Academic Credentials 268
13.3 Exaggeration/Overstatement of Academic Track Record .. 271
13.4 Exaggeration/Overstatement of Research Findings 273
13.5 Falsification/Fabrication of Results 276
13.6 Plagiarism and Failure to Acknowledge Work 278
13.7 Theft of Intellectual Property (IP) 280
13.8 Misappropriation/Misuse of Research Funds 281
13.9 Other Areas of Misfeasance 283

14	**The Long Term Supervisor/Researcher Relationship**	**285**
	14.1 Overview	285
	14.2 Academic Career	288
	14.3 Professional Career in Broad Field of Research	290
	14.4 Management Career	293
	14.5 Commercial Research and Development Career	295
	14.6 Start-Up Company	297
	14.7 Other Basic Relationship Building Issues	300
15	**Questions Supervisors Should Ask Themselves**	**303**
	15.1 Overview	303
	15.2 What Is My Motivation for Research Supervision?	303
	15.3 How Will My Personal Ambition Impact on Supervision?	304
	15.4 How Do I View My Role in the Research World?	304
	15.5 What Is My View on Research Supervision?	305
	15.6 What Sort of Professional Traits Do I Have?	305
	15.7 Am I Currently in a Good Position to Supervise a Student?	306
	15.8 The Most Important Question	307

Appendix   309

Index   313

About the Author   315

Foreword

Henry Ward Beecher once wrote that,

"We should not judge people by their peak of excellence; but by the distance they have traveled from the point where they started."

In this context, the supervision of a postgraduate research candidate should be a journey of knowledge for both the research candidate and the supervisor. However, for the supervisor, the journey is far more complex than the conduct of the research itself. It is ultimately about assuming a level of responsibility for a human being – more specifically, their physical and emotional safety, wellbeing, learning and professional development. Unlike the undergraduate learning environment, where one academic seeks to educate many students simultaneously, in postgraduate research supervision, there is a one-to-one relationship, which is intrinsically more difficult to manage.

Some academics believe that the process of research supervision is centered around a professional bond between a supervisor and student, a shared interest in an area of research, a common desire to achieve field-specific outcomes, and provision of mentorship to the student. This may be partly or wholly true, but it is an oversimplification. A modern university is not just a place of learning, it is also a workplace – and a complex one at that. Each institution has numerous internal and external regulations, as well as processes that must be addressed. Beyond these, there exist national and regional laws that impinge upon the supervisory role, in terms of treatment of students and staff, as well as general research ethics issues.

When disputes arise, both the institution and the supervisor need to ensure that the supervisory process has been undertaken in a diligent, systematic and professional manner – and not solely based on an informal/friendly relationship between supervisor and student. The responsibilities of supervision are therefore significant, with potential career ramifications and legal sanctions, in addition to basic academic objectives.

A common challenge that academics face in tackling new areas of responsibility is an overload of information and ideas. This is particularly true in the case of research supervision. For many activities there are innumerable valid approaches to the tasks at hand. Reading and understanding all these approaches can become an insurmountable challenge in its own right – and an impractical one, given the demands on academic/research time. In addition to the various approaches, one then has to superimpose consideration of all the specific rules and procedures imposed by the universities themselves. How then is one to tackle the complex issue of postgraduate supervision for the novice supervisor?

I have found that the best approach to tackling a complex task in a complex environment is to evaluate what an experienced practitioner does, and then formulate an approach around that – borrowing, adapting or rejecting elements as required – in order to achieve something which suits the specifics of the task at hand.

In 1999, I released the first edition of *Key Factors in Postgraduate Research – A Guide for Students*. In that book, I took the decision to present the issues in a straightforward, linear manner, so that students could read the book from cover to cover in a few hours, and then decide on what additional information they required, or what alternative approaches they could take. Since that time, many thousands of copies of the book have gone into circulation at universities around the world, and the feedback has been positive. People liked the idea of a guide that could be read as a regular book, rather than an academic treatise that covered every conceivable angle. Many research students have used the book as a background template, in order to develop their own approach.

The view taken herein, therefore, is similar. Rather than presenting potential supervisors with every conceivable approach to their task, the aim is to present a broad, linear perspective, and then to allow people to use this as either a point of reference or a point of departure. The assumption is that those entrusted with the supervision of a postgraduate researcher have the judgment/wisdom to formulate their own specific approach – or, perhaps, explore an array of other opinions when they encounter something herein which doesn't sit well with them.

I have supervised numerous postgraduate research students to successful completion and, more importantly, have made sufficient errors of judgment to learn the hard lessons of what happens when things do go wrong. I have also been called in on numerous occasions to assist in resolving problems that have arisen when the relationship between a supervisor and student has broken down. The broad lesson is that any competent academic can be a good research supervisor when things are going well – real supervisory skill only comes to the fore when a supervisor/student relationship isn't working, or the research program itself has gone completely awry.

A key lesson I have learned after more than two decades of supervision, and working as a problem solver and intermediary, to resolve other supervisions that have gone awry, is that no two supervisions are the same; no two supervisors and the same, and no two students are the same. It would therefore be simplistic to believe that there is a single, universal model that works. Further, no amount of reading will be sufficient to resolve all the problems that supervisors will encounter. Practical experience, mistakes and, most importantly, learning from those mistakes and hopefully not repeating them, are ultimately the only real learning tools that we have.

This book cannot realistically attempt to prevent supervisors from making errors of judgment. However, by providing a broad background to the supervisory process and the common errors that occur – as well as their resolution – supervisors may be able to preempt and correct problems before they become serious.

Notwithstanding these realities, there is one cardinal rule that does need to be taken seriously – and that is that every decision that a research supervisor makes has to be in the best interests of the research student. If this rule is observed in full, then many of the other problems that will inevitably arise during the course of the supervisory process can be resolved in a professional manner.

Dr. Dario Toncich

Acknowledgments

I would like to thank sincerely Dr. Ben Inglis, of the Henry H. Wheeler Jr. Brain Imaging Center at the University of California Berkeley, for his contributions to proof-reading this book and for his many thoughtful insights into its contents and structure.

List of Figures

Figure 2.1	Pasteur's Quadrant Diagram for Research (Stokes, 1997).	24
Figure 2.2	Understanding the basic problem of research writing.	29
Figure 4.1	TKI conflict handling modes.	92
Figure 5.1	Basic elements of the modern university.	103
Figure 5.2	Typical elements in modern universities.	106
Figure 5.3	Typical university governance structure.	107
Figure 8.1	Basic elements of project management chart.	152
Figure 12.1	Complexity of argument flow in a thesis.	227
Figure 12.2	Conceptualizing the literature review funnel.	243
Figure 12.3	Chapter linkages in a thesis.	254
Figure 12.4	Bringing individual thesis elements together to create a compelling concluding chapter for the thesis.	258

List of Tables

Table 1.1	Generic sequence of steps in the postgraduate research supervisory process	12
Table 2.1	Outcomes for research students and supervisors	16
Table 2.2	Research student skills development outcomes	19
Table 2.3	Basic steps in structured postgraduate research	21
Table 2.4	An example of a sentence by sentence parsing tool to check writing discipline	33
Table 2.5	Example oral/visual presentation template	38
Table 2.6	Example of a thesis structure	48
Table 3.1	Developing a health/safety risk assessment checklist for research students	59
Table 3.2	Example of possible development pathways and requirements	75
Table 4.1	Common supervisor beliefs	82
Table 4.2	Possible advantages and disadvantages of supervisor types	83
Table 4.3	Typical supervisor/student disputes (student perspective)	88
Table 5.1	ARWU ranking criteria (Shanghairanking.com, 2015b)	99
Table 6.1	Example of costing for postgraduate research project	116
Table 6.2	Candidate attributes that supervisors need to consider	123
Table 7.1	Sample template for meetings between supervisor and research student	140
Table 8.1	Basic elements and time considerations is postgraduate research	149
Table 9.1	Stages of conflict in disputes	157
Table 10.1	Sample project milestone/deliverables chart	187

Table 10.2	Differences between academic and commercial imperatives	201
Table 10.3	Pro-active tasks for supervisor to ensure student welfare	212
Table 12.1	Basic seven chapter thesis template	225
Table 12.2	Thesis preparation time considerations example	234
Table 12.3	Sample template of a literature review chapter	246
Table 12.4	Balancing between thesis body and appendices	248
Table 12.5	Commonly used words/phrases which create issues in theses	251
Table 13.1	Steps for dealing with misfeasance	267
Table 14.1	Example of possible development pathways and requirements	287

List of Abbreviations

arXiv	Archive Research Preprint Server (Cornell University)
bioRxiv	Bioscience Archive Research Preprint Server (Cold Spring Harbor Laboratory)
CEO	Chief Executive Officer
CITI	Collaborative Institutional Training Initiative (University of Miami)
CSIC	Consejo Superior de Investigaciones Cientificas
CWTS	Centre for Science and Technology Studies (Leiden University)
IELTS	International English Language Testing System
IP	Intellectual Property
IT	Information Technology
NIH	National Institutes of Health (USA)
NSF	National Science Foundation (USA)
PhD	Doctor of Philosophy
QS	Quacquarelli Symonds
R&D	Research and development
THE	Times Higher Education
TOEFL	Test of English as a Foreign Language

PART I

Preparation

1

Introduction and Text Usage Guide

1.1 Aims and Text Structure

In 1945, in his report to US President Harry Truman – *Science: The Endless Frontier (Nsf.gov, 2015)* – Vannevar Bush, Director of the US Office of Scientific Research and Development, wrote the following words:

> *"The responsibility for the creation of new scientific knowledge – and for most of its application – rests on that small body of men and women who understand the fundamental laws of nature and are skilled in the techniques of scientific research. We shall have rapid or slow advance on any scientific frontier depending on the number of highly qualified and trained scientists exploring it."*

Decades later, we are still dependent upon the skills of highly qualified and trained scientists for our societal advancement. This book relates specifically to those who are involved in the training of researchers – not just scientists, but all those who practice research in diverse fields.

The objective of this book is to provide a framework guide to the key issues that arise during the conduct of postgraduate research student supervision. This book is aimed primarily at early career academics and researchers who may be tackling the supervisory process for the first time, either as a principal or associate supervisor in a postgraduate degree program.

The specific structure and assessment of postgraduate research degree programs varies from country to country and institution to institution. In some institutions, a Master's degree (by research) is a distinct research program in its own right – in other institutions it is a precursor to a Doctoral program. For some institutions, a postgraduate research program is composed entirely of research into one specific area – in others, research is supplemented by coursework studies.

The complete picture is also remarkably variable at an international level. In the United States, there are often qualifying examinations for entry to higher level postgraduate programs such as Doctorates. Supervisors sometimes act as a leading entity in a thesis committee rather than unilaterally in their own right. In the British Commonwealth universities, supervision is generally left to an individual – or perhaps a principal supervisor and a co-supervisor.

Assessment methods also vary significantly, with some institutions requiring only the peer evaluation of a dissertation/thesis, and others expecting a dissertation supplemented with a formal defense of the work.

Even within a single institution, there can be marked variations to postgraduate research program formats – for example, some universities offer special forms of Doctorate, based upon a collection/portfolio of published work. Industry-based Doctorates add further variations, with the possibility of including commercial research and development into a program of assessment.

The possible postgraduate program variations are clearly significant and it is not practical to cover all of them in detail within a single book. Hence, herein, the entire suite of possible postgraduate research programs is condensed into the generic context of a single research program that is ultimately peer-assessed by means of dissertation and/or verbal defense. There is also an assumption that, for the bulk of the research program, there will be a professional relationship between a supervisor and student, in which that supervisor takes responsibility for oversight of the research conduct and preparation of research papers, thesis and defense.

This book is written in such a way that it can be read within a few sittings, in order to provide an overall insight into the basic aspects of research supervision – from initial contact with a research student, through to final thesis submission. Importantly, this book is not intended to be a research text in its own right – with large numbers of references and citations. The objective is to provide an easily digestible, linear pathway, covering the supervisory process, from beginning to end, in a straightforward manner – and written in the form of a guide.

Albert Einstein once observed that,

"Everything must be made as simple as possible. But not simpler."

With this in mind, the context of this book is important for the reader to understand. It begins with the premise that the process of research supervision is actually complex in its own right. Add to this the fact that each university has specific guidelines for postgraduate research conduct and examination. Add to this the fact that individual faculties, departments, research institutes, centers and research groups have their own research guidelines, under the umbrella of the university structure. Add to this the fact that some postgraduate research programs will involve external partners – including business/industry as well as external research organizations. Add to this the different capabilities of research students, and their varying requirements for support and interaction with their supervisors. Again, the permutations are enormous.

It would be neither helpful nor feasible to address all the permutations of research environments and programs, and then expect a reader to extract specifically relevant supervisory practices from such a text. Moreover, it would be unwieldy to assemble such a range of perspectives into a cohesive collection of methods that an individual supervisor could use practically. Instead, this book is written in such a way as to provide a sequential perspective on each of the key aspects of the supervisory process. These perspectives can then act as a point of reference – or a point of departure – for the reader.

It would also be presumptuous to suggest that what is presented herein is an optimal approach to supervision or tackling the issues intrinsic to that process. Ultimately, the best approach to supervision will arise from the supervisor's understanding and responses to:

- The surrounding research environment and resources.
- The specific field of research, and peer expectations within that field.
- The specific attributes – strengths and limitations – of the research candidate.
- The developing professional relationship between the research candidate and the supervisor.
- The surrounding peer groups available for consultation and support for both the supervisor and the student.
- External factors influencing the research program, including collaborating partner organizations, etc.

A research supervision undertaken in one of the world's highly-resourced and highly-ranked universities may be markedly different to one undertaken in a resource-poor regional university in a developing nation. This doesn't have to mean that the quality of the outcomes will be different, but it does mean that supervisors need to adapt and supervise according to their surroundings.

There are as many approaches to research supervision as there are research supervisors. It would therefore be condescending to supervisors to become too prescriptive in the approaches that such individuals should take. Herein, the objective is to provide a basic, common approach to tackling each key supervisory issue, and to work on the assumption that the readers will then calibrate their specific approach accordingly, based upon their own professional judgment.

The basic elements of postgraduate research supervision that need to be addressed include the following:

(i) Identifying/selecting/evaluating appropriate candidates for research, and initiating discussions with them.
(ii) During early discussions, identifying potential career pathways for the candidate at the completion of the research program, and determining how the program can be structured to maximize opportunities for the candidate to achieve his/her goals.
(iii) Defining the specific terms of the research candidature and project for the benefit of the researcher, supervisor and institution.
(iv) Establishing the boundaries for a professional relationship between the supervisor and research candidate, and articulating how that relationship will manifest itself over the research period.
(v) Identifying potential problems that may arise during the course of research candidature – understanding potential personal problems that a candidate may have; potential health and safety issues with the research to be performed; resource requirements, and then attempting to identify the unique requirements of the candidate.
(vi) Planning the research program and developing a mutually agreed project management plan and timeline with the candidate.
(vii) Establishing regular, formal and informal, meetings during the course of the research program in order to provide constructive, ongoing feedback in relation to the research program and its directions.
(viii) Determining internal and external peer review mechanisms throughout the course of the research to ensure integrity – including discussions with local peers, learned colleagues; publication of findings, etc.
(ix) Formally supervising – and providing oversight to – the research/ experimental work of the candidate to ensure that the candidate is performing according to agreed requirements.
(x) Ensuring that the research candidate is behaving morally, ethically, and adhering to the rules of the institution.
(xi) Ensuring that the research candidate acts in a safe manner that does not imperil himself/herself or others working in the environment.
(xii) Ensuring that the candidate behaves in a manner which is respectful and courteous to colleagues and supporting staff.

(xiii) Providing ongoing advice to the candidate about achievements and what areas the candidate needs to improve in order to achieve his/her goals beyond the program.
(xiv) Identifying problems that the candidate is unable to resolve in his/her own right, and determining what additional support measures and resources may need to be introduced in order to complete the research program.
(xv) Resolving ongoing problems as they arise – whether they be personal; research-specific; resource-based or professional conflicts with other students/staff.
(xvi) Providing ongoing support and feedback in relation to the preparation of research papers and the thesis/dissertation.
(xvii) Preparing the candidate for examination by thesis and/or other defense.
(xviii) Discussing future plans and career prospects with the candidate.
(xix) Establishing an ongoing professional relationship with the candidate beyond the research candidature.

This list is not exhaustive, but it does serve to demonstrate that research supervision is far more complex and multidimensional than the basic oversight of a set of research activities.

Novice research supervisors can sometimes naively believe that if they are exemplary mentors and loyal friends to their research students, then all else is peripheral, and any other problems will resolve themselves. The simple fact of the matter is that they will not. While research supervisors may indeed be exemplary friends and mentors to their students, they have a professional responsibility to their university to manage their researchers in a systematic, methodical and impartial manner – and according to university procedures. This duty of care needs to remain intact whether or not a supervisor and student are friends, and even if a student does not perceive the supervisor to be one of his/her mentors.

At the most fundamental level, each research supervisor needs to understand that he/she is assuming an important level of responsibility for a human being – specifically, one that will need to interact with other human beings, and potentially operate in an environment that may contain various hazards. Additionally, the candidate may be required to perform tasks which are potentially deleterious to other humans or animals used in experimentation. These are not minor, sundry issues – they are issues that have profound moral and legal consequences in most modern, developed countries.

Over and above these core responsibilities for the supervisor, there needs to be an ingrained awareness that the postgraduate research program is creating a highly specialized individual – and one for whom career opportunities will, by definition, be limited accordingly. This makes the responsibilities of the supervisor all the more complicated.

At a global level, each year, the number of individuals completing postgraduate research qualifications is significantly higher than the number of available tenured academic positions within the *research-intensive* university sector. Moreover, the deficit between graduates and tenured academic positions is annually cumulative. It would therefore be reasonable to suggest that many of those who complete a postgraduate research qualification will *not* be working within the university system as an academic in the long-term. Some will work in commercial research; some will work in normal professional roles in commercial or government entities; some will move directly into management roles, and some will create their own start-up companies.

Whichever pathway a postgraduate researcher ultimately chooses, there needs to be recompense for opportunity costs and forgone professional income as a result of participation in a postgraduate program that takes several years to complete. It needs to be recognized that this period of postgraduate research is 5–10% of a professional working life, and so the compensation needs to be significant. This compensation may be monetary but it could also come in the form of more interesting work; a business opportunity; better overall career prospects, or just some form of personal satisfaction as a result of achieving a particular research outcome.

For these reasons, the actual research which students undertake and complete during their time as a postgraduate cannot be *the* outcome – it can only be one part of the broader set of outcomes that need to be achieved if the university and supervisor are carrying out their roles diligently. In the modern world, a good piece of research and an unemployable graduate is as inadequate an outcome as a poor piece of research and an employed graduate.

A supervisor therefore has an important role in achieving both a good research outcome and a postgraduate with meaningful future professional directions. Without this, what does any university really achieve in the long-term – if its graduates are not deemed to be employable in positions befitting their highly qualified status?

Unfortunately, sometimes universities and academics view postgraduate research students as little more than low-cost research labor, in order to achieve their own ends. However, this is a very shortsighted and dated view – particularly if the end result is a collection of disenfranchised, disgruntled individuals with limited career prospects. Taking the longer term view, and understanding that successful, grateful graduates will inevitably give back immeasurably to the university and its research is a far more productive approach for the modern world.

For these reasons, creating a *complete postgraduate* should be the ultimate goal of the modern supervisor, and it is a difficult one. In order to achieve this, research students will need to be pushed outside their comfort zones. However, this means that research supervisors will also need to step outside their own comfort zones and look at the broader development of the complete high caliber postgraduate as part of their core responsibility. This may seem unjust, given that supervisors can view themselves solely as purveyors of knowledge in a specific field, but it is the reality of modern supervision.

In many universities, the guidelines for supervisors talk of *mentoring* research students. The most basic part of that mentoring role is setting a good example in the context of research conduct and, more particularly, in terms of research ethics. Another obvious part of the mentoring role is trying to get research students to understand their own strengths and limitations. In the modern world, however, a new and critical aspect of supervision is getting research students to understand the varied and complex environments into which they will ultimately transition at the conclusion of their postgraduate degree. This is not something which can be achieved within a few weeks at the conclusion of the program – it is something which needs to start early in the research candidature.

Completing a significant research task is one part of modern postgraduate candidature, and understanding how that research task, and the broader research training, fit into the world is the other.

On reaching the end of this text, and given the range of supporting supervisory activities which become apparent herein, novice supervisors may be forgiven for thinking that all these supporting activities will detract from the actual journey of knowledge discovery. This is not the case in practice. As a supervisor develops increasing experience in the oversight of postgraduate research, the supporting activities will become second nature, background thought processes, and the primary focus will always remain on the discovery of knowledge.

1.2 Text Overview

This book is composed of three main parts:
  I. Preparation
  II. The Supervisory Process
  III. Relevant Supervisory Issues.

The first part of the text deals with issues that relate to preparing for and initiating the supervisory process. The second part deals with issues directly relating to the research supervision and examination itself. The third part deals with a range of supporting issues that are not core to the supervisory process itself but can have critical ramifications for the conduct of the research supervision and the future career of the candidate.

The three parts of the book are divided into a total of 15 chapters, 14 of which follow on from this one. In summary, the purpose of each of the subsequent chapters is as follows:

PART I – Preparation:

- *Chapter 2 – Understanding the Fundamentals of Research Supervision* – A chapter looking at the critical outcomes that need to be delivered for both the student and the supervisor
- *Chapter 3 – Basic Responsibilities of Research Supervision* – A chapter outlining important basic requirements for supervisors, including maintaining safety and ensuring wellbeing (physical and emotional) for students – including avoidance of discriminatory practices, etc.
- *Chapter 4 – The Relationship between the Supervisor and Student* – A chapter examining the types of supervisory relationships that can exist and the relative benefits/limitations.
- *Chapter 5 – Understanding the Research Environment* – A chapter explaining the different types of research environments – generic governance structures in universities, research institutes, etc.

PART II – The Supervisory Process:

- *Chapter 6 – Preliminary Tasks in Research Supervision* – Student recruitment; definition of research projects; scholarship funding; submission of candidature forms, etc.
- *Chapter 7 – Initiating the Formal Supervisory Process* – Establishment of formal and informal meeting mechanisms; project requirements, and so on.
- *Chapter 8 – Planning the Postgraduate Research Program* – Regardless of discipline or research topic, all research programs have basic elements, including literature review; development of methodology; experimental design or hypothesis testing; analysis of results; peer-review by publication; preparation and presentation of a thesis – and possibly a verbal defense. These need to be assembled into a cohesive plan.

- *Chapter 9 – Conflict Resolution in Postgraduate Research* – There are numerous sources of conflict that arise in postgraduate research – including those between the supervisor and student; the student and colleagues; resource providers; external parties to the research, and so on. A supervisor needs to be able to identify the sources of these conflicts and put in place mechanisms for their resolution.
- *Chapter 10 – Research Supervision in Industry/Partnered Collaborative Projects* – When research is conducted in collaboration with third parties (either commercial or other universities and research institutes) additional considerations need to be factored into supervisory practices. In particular, these can include contractual disputes; conflicts of interest; conflicting objectives; intellectual property disputes; publication disputes, etc.
- *Chapter 11 – Peer Review by Publication* – In most postgraduate research programs, a preliminary part of the research assessment occurs by having the student submit research papers on key elements of the work. This chapter deals with issues that arise as a result of the publication process.
- *Chapter 12 – Preparation of a Thesis* – Postgraduate research programs are structured investigations, and subsequently theses also have to be carefully structured in order to document those investigations. This chapter looks at the generic elements that need to be included, and the various options for thesis structuring.

PART III – Relevant Supervisory Issues:
- *Chapter 13 – Research Misfeasance Issues* – In an ideal world, research misfeasance should not arise but it unfortunately does during the course of an academic career. This chapter looks at generic issues relating to plagiarism or other academic misconduct on the part of student, supervisor or surrounding academic colleagues.
- *Chapter 14 – The Long Term Researcher/Supervisor Relationship* – In order to help build a vibrant university, supervisors need to ensure that the end of the postgraduate research program is not the end of the relationship between the student and supervisor. In the short term, supervisors may need to become involved in supporting the career choices of their students. In the long term, there is a need to ensure an ongoing positive relationship.
- *Chapter 15 – Questions Supervisors Should Ask Themselves* – Having read the preceding 14 chapters, and before actually taking on a research student, potential supervisors need to ask themselves some basic questions about their motivations and how these will impact upon a candidate.

1.3 Overview of the Research Supervisory Process

In order to progress discussions on the various aspects of research supervision, some assumptions need to be made about the sequence of events associated with research supervision. Needless to say, the specific elements in the process, and their sequencing, are unique to each particular university and each field of research. However, in order to make the discussions herein tractable, a generic sequence of supervisory elements is presented in Table 1.1, as the basis for discussions herein.

Table 1.1 Generic sequence of steps in the postgraduate research supervisory process

Step	Details
1	Identification/selection of postgraduate research candidate
2	Preliminary/informal discussions with potential candidate to establish possible research areas/projects, and long-term career aspirations
3	Detailed discussions with potential candidate to establish research program specifics and funding/scholarship/tuition arrangements
4	Formal application and university processing of research candidature application and scholarship/tuition funding arrangements
5	Initial meeting/s with candidate to formally commence research program and set off a series of regular supervisor meetings. Formal/informal discussion of supervisory arrangements and general expectations on the part of the supervisor and candidate
6	Formal establishment of basic resources for research candidature – including office and laboratory space; technical; library and information technology (IT) support, etc.
7	Identification and evaluation of potential hazards; safety requirements, and ethics requirements for the research program
8	Initial research candidature project meeting – outline of project requirements and project plan; timelines; critical pathways; thesis preparation process; interim and final milestones/deliverables; etc.
9*	Regular formal/informal meetings to discuss and evaluate research progress; results problems and resolution mechanisms
10*	Meetings to discuss formal and informal internal/external peer review processes for the research candidate – including in-house presentations of interim work; reviewed publication of results, etc.
11*	Meetings to discuss future professional pathways for candidate at the completion of the postgraduate research program, and possible mechanisms for achieving these
12*	Meetings to discuss thesis preparation/progress and final examination of research where verbal defense is required

1.3 Overview of the Research Supervisory Process

13*	Problem resolution meetings to address personal and professional research student issues; conflicts; health or safety concerns
14*	Editing, feedback and iterative development of research dissertation/thesis
15	Preparation of candidate for final thesis submission and/or examination
16	Meetings to discuss future career directions and employment possibilities for candidate – support for research candidate's transition out of the program
17	Establishment of a future, ongoing professional relationship with the candidate to enable the professional researcher to remain an active contributor to the university environment.

*Sequence steps marked with an asterisk are repetitive, ongoing elements of the process.

2

Understanding the Fundamental Outcomes of Research Supervision

2.1 Overview

The purpose of this chapter is to provide an insight into what the supervisor and the research student might seek to achieve as outcomes during the course of a postgraduate research program. This chapter therefore provides an overview of a broad range of issues, many of which will also be covered again, in more detail or from a different perspective, in subsequent chapters.

Those tasked with the supervision of a postgraduate research student will generally have already completed their own graduate research program and will be acutely aware of the basic requirements – that is,

- The completion of a systematically conducted program of research investigation.
- Contributions to the relevant field of knowledge.

However, these are only part of what needs to be achieved in order for the research candidature to be deemed successful, in the context of adding real value to the individuals involved. The others are intrinsic to the conduct of the research itself and to the development of the postgraduate student as a high caliber professional.

Table 2.1 provides a summary of the areas that will be covered in this chapter.

Two elements that aren't covered in Table 2.1, but which form critical learning outcomes from the postgraduate research learning process, are:

- A recognition that an important part of any research process, which seeks to extend knowledge, often involves the making of mistakes – getting things wrong – recognizing the errors, and moving forward from that point.

- Differentiating between what it is actually known; what is factual, and what is mere opinion or conjecture.

These may seem rudimentary to an experienced researcher but not so to recent graduates who have spent almost two decades of rote-learning during their school and university education.

Specifically, research students tend to come into the research environment from a paradigm where *correct answers* are rewarded and *incorrect answers* are penalized. They also come from a learning environment where an admission of *not knowing* is often a taboo. In some undergraduate learning environments, there is a failure to teach students that published literature is not infallible – that there can be errors or omissions, or biases – and often published material can be superseded – or rendered invalid – by changing

Table 2.1   Outcomes for research students and supervisors

Outcome	Student	Supervisor
Development of fundamental research skills (literature review, experimental design, presentation and defense of hypotheses, etc.)	✓	
Execution of a structured program of research	✓	
A contribution to (extension of) knowledge in a particular field	✓	✓
Development of rigorous documentation techniques and disciplined writing style	✓	
Development of oral/visual research presentation skills	✓	
Peer evaluation/acceptance of research	✓	✓
Published research findings	✓	✓
An understanding of the broader perspective of the research and research methods – and where these potentially fit into a graduate's ongoing career expectations	✓	✓
Establishment of a high caliber research team in which the research student is an active and valued member		✓
Creation and maintenance of a dynamic research and learning environment which actively encourages the exchange of ideas and opinions		✓
Establishment of a research team in which peer review of each member's work is an intrinsic part of the research effort.		✓
A dissertation/thesis for examination and defense	✓	
A professional career foundation	✓	
Underpinning research outcomes for future research grants		✓
Research outputs that feed into university research performance metrics (publications, citations, Doctoral completions)		✓
Establishment of a long-term professional relationship between the graduated student and the supervisor	✓	✓

knowledge. In the context of international students, it is also important to be aware that in some cultures, challenging or even questioning orthodoxies is discouraged in the undergraduate environment.

Postgraduate research students therefore need to come to terms with the idea that errors and mistakes are potentially a positive because they are a basic learning tool in research. An admission of *not knowing* is a hallmark of honesty rather than learning deficiency. To quote from George Washington,

> *"...I trust the experience of error will enable us to act better in future."*

This recognition needs to apply equally to a postgraduate research supervisor. Each supervisor eventually recognizes that, in practice, the supervisory process rarely turns out to be a simple, linear sequence of steps from student induction to graduation ceremony. Specifically,

- Mistakes will be made.
- Setbacks are altogether common.
- Disputes will occur.
- Resources will fall short of expectations.
- Some students will drop out of their research programs and blame their supervisor.
- Some students will lodge formal complaints against their supervisors for one reason or another.
- Some students will fail during the examination/defense portion of the postgraduate research assessment.

These negative aspects therefore also need to be considered as possible outcomes of the research supervision process. Whether any or all of these occur during the very first supervision is partly a question of serendipity, and partly a question of how much effort and consideration a supervisor puts into a research program.

Many novice supervisors commence their first supervision with high hopes and expectations, and considerable enthusiasm. It is not until the end of the research program is in sight that they recognize the scale of their responsibility – they are ultimately responsible for the future of a human being and for the consequences, if that human being fails to achieve his/her research and career objectives. Better then to consider what is required at the outset, than to be responsible for negative outcomes that could cause considerable personal, emotional and professional damage to a postgraduate research student's life at the end of the process.

2.2 Development of Fundamental Research Skills

The driving forces behind postgraduate research supervision are important to understand. For some research supervisors, the specific research project and its completion are the only objectives of the postgraduate study program. For others, who may hold a higher sense of purpose for their research students, there is a more mature recognition that the research project is predominantly a vehicle that is used in order to enable the research student to develop specific professional research skills.

At the highest levels of postgraduate research – the Doctorate – the program is fundamentally a vehicle to learn the skills of independence, no matter the subject, the starting point or the goal. If a program is structured correctly, a postgraduate should ultimately be able to:

- Locate, read and review pertinent literature.
- Establish a program of investigation.
- Document progress.
- Meet research targets or milestones.

That one has a background in, say, chemistry should not preclude those skills being applied to any other field, even a non-scientific one. It is independence, structured analysis and execution that sets a high level postgraduate research degree apart from any other qualification.

Table 2.2 shows the sorts of skills that supervisors need to work towards developing in their research students. There is a hierarchy – commencing with the most basic ones, pertaining to the conduct of a professional program of research investigation. At the next level, supervisors need to consider the development of their research students into mature research professionals – ones who are prepared to act in a completely impartial and ethical manner. Finally, there is a need to instill a level of personal, professional integrity and maturity within a research student, if it does not already exist at the commencement of the program. Perhaps the most important among these traits is a willingness, on the part of the graduating student, to take on the important and lifelong role of being a gatekeeper of knowledge – that is, a person who values the sanctity of knowledge above personal self-interest.

Not all research students will be capable of achieving all these lofty ideals but they should serve as a benchmark for supervisors to work towards.

2.2 Development of Fundamental Research Skills

Table 2.2 Research student skills development outcomes

Level	Description	Outcomes
1	Basic Skills	An ability to conduct a comprehensive literature review.
		Based upon the outcomes of the literature review, and identified knowledge gaps therein, an ability to systematically develop a hypothesis which needs to be investigated rigorously.
		An ability to systematically investigate and consider the ethical implications of investigating the hypothesis – particularly where a hypothesis may pertain to humans or animals.
		An ability to develop/formulate a methodology that can be used to explore a hypothesis, within the ethics guidelines of the university environment.
		An ability to develop instruments (experiments, models, surveys) that can provide factual evidence to support or contradict the hypothesis.
		An ability to systematically aggregate and analyze data in order to determine the efficacy of the hypothesis.
		An ability to draw specific conclusions based upon the factual information derived from the methodology.
2	Professional Research Skills	A capacity to assess the health and safety implications of a research methodology and experimentation program.
		An ability to conduct one's research to an *open hypothesis* that does not predispose the research methodology and results to a particular outcome.
		An ability to act, at all times, as an impartial observer that presents a balanced view of the work of others as well as his/her own work.
		An ability to understand the broader implications and value of the contributions and value of one's research – within the research field and within the commercial/societal context.
		An ability to understand the context and relative contributions of one's own work in the broader field of endeavor – for example, is the postgraduate research an insignificant input in the context of an area that has already had several thousand person-years of effort already expended within it?
3	Personal Professional Integrity	An ability for critical self evaluation – the willingness to openly present and compare the worst possible interpretation of one's own work against the best possible interpretation of others' work.
		A capacity for humility/modesty in relation to one's own research – an ability to avoid overstatement and to allow the facts to speak for themselves.
		An ability to function in a team environment where individual strengths, differences and weaknesses are used to create a whole which is greater than the sum of its parts.
		An ability to put to one side one's own personal and career objectives in the search for knowledge – more specifically, a preparedness to put one's career on the line when the ethics of research or integrity of knowledge are threatened.

2.3 Execution of a Structured Program of Research

The more that one supervises postgraduate research projects, the more that one recognizes the similarities between them – particularly in terms of structure. Universities each have their own specific requirements for postgraduate research degree programs and assessments, but the fundamental elements tend to transcend institutions and fields of research. Table 2.3 shows the common elements that need to be achieved within a structured program of research.

Most of the elements in the table will already be known to those entrusted with the task of research supervision but it is necessary to highlight the importance of a few points therein.

Firstly, research students need to learn and understand that a systematic program of investigation is not based upon a *thought bubble* of what a student thinks might be a good idea in order to create new knowledge. The structured investigation needs to fit into a sequence of peer research that may have been under way for years, decades or centuries – and it needs to mesh with the views of learned peers in respect of where knowledge gaps exist.

Many research students, in their youthful enthusiasm, long to be paradigm-shifters, but the reality of postgraduate research is that they need to start their work as *incrementalists*, seeking to logically extend knowledge in gaps identified by other learned scholars. For this reason the literature review is the critical foundation for the research.

Secondly, students need to come to terms with the concept of the *null hypothesis,* and how this should fit in with their research objectives. In the narrow statistical sense, the Merriam Webster Dictionary *(Merriam-webster.com, 2015)* defines a null hypothesis as follows:

> *"...a statistical hypothesis to be tested and accepted or rejected in favor of an alternative; specifically: the hypothesis that an observed difference (as between the means of two samples) is due to chance alone and not due to a systematic cause."*

In other words, statistically, the null hypothesis of a postgraduate research program is that two sets of data are the basically the same and that differences do not have an identifiable, underlying cause. In a broader sense, the null hypothesis in research is that whatever a research student seeks to exert in his/her research has no effect on the entity being tested.

2.3 Execution of a Structured Program of Research

Table 2.3 Basic steps in structured postgraduate research

Step	Function	Description
1	Literature Review	The literature review provides background historical and current information on the field of interest. It identifies seminal findings and events; key researchers in the field; gaps in knowledge, controversies and disputes in knowledge and approaches, and potential future research directions. In structured research programs, a research student needs to demonstrate that his/her research hypothesis and directions from research arise from systematic reviews of the work of learned peers in the field, rather than as random ideas.
2	Formulation of Hypothesis	The research hypothesis is derived from a formal understanding of the gaps in knowledge which have been identified as a consequence of the literature review. The key objective is to ensure that the hypothesis is *open* and does not predispose/skew the research to a particular outcome.
3	Development of a Methodology	The methodology is the sequence of steps taken in order to substantiate the reasons for putting forward the hypothesis and then testing that hypothesis in a systematic and impartial manner, without necessarily having cognizance of the end result.
4	Design of Instruments/ Experiments to Evaluate Hypothesis	A series of instruments/experiments needs to be developed systematically in order to formally determine the validity of the hypothesis. Typically, these instruments/experiments would need to conform with normally accepted practice in the particular field – and the basis for them may need to be substantiated through the citation of independent, published scholarly work.
5	Analysis of Results/Findings	A systematic and fair (impartial) analysis of the findings needs to be presented to address the hypothesis.
6	Local Peer Review of Results/Findings	Research findings/results need to be presented for peer evaluation at a local level, within the center, institute, department or faculty to validate or repudiate the outcomes.
7	External Peer Review/ Publication	Once a local peer review is completed, the research findings need to be presented to a larger audience – either through publication in scholarly, reviewed journals or conferences, in order to achieve independent validation/repudiation of outcomes.
8	Evaluation of Strengths/ Limitations of Research and Results/Findings	The research student's project is generally only one small element in a chain of research in a given field. The research student needs to critically self-evaluate his/her research and identify shortcomings, weaknesses, etc. as well as strengths. The significance of the research student's contribution within the much larger chain of research in the field also needs to be determined.
9	Documentation of Research Program and Findings	The research work needs to be presented within a detailed document – dissertation/thesis – for assessment and defense as well as for the purposes of creating a historical record of the research and findings.

The research student's role is therefore to *disprove the null hypothesis*, and thereby establish that his/her contributions have a genuine impact on the entity being examined.

In the context of the null hypothesis, it needs to be noted that the formulation of an *open research hypothesis* is also particularly important to the conduct of a systematic and impartial investigation. An *open hypothesis* is one which does not predispose the research to a particular outcome but merely seeks to determine, in an impartial way, whether or not a particular phenomenon occurs. For example, an *open hypothesis* might be as follows:

"The objective of this research was to determine whether or not an increase in the administration of Xylatrin would alter the incidence of strokes in patients already using hypertensive medication."

A *closed hypothesis* might be as follows:

"The objective of this research was to demonstrate that an increase in the administration of Xylatrin would decrease the incidence of strokes in patients already using hypertensive medication."

Note that in the *open hypothesis*, the research produces a useful outcome, regardless of the experimental results. In the *closed hypothesis*, the research seemingly fails if it does not demonstrate a predefined outcome. In other words, it tends to skew the thinking of the researcher – and it dissuades that researcher from being an impartial observer in the research program. This is a fundamental flaw that needs to be picked up by the supervisor at the outset of the research, in order to ensure the integrity of the program of investigation.

The other point that needs to be made about the steps in Table 2.3 relates to local peer review. The most important review that a research student will receive for his/her research needs to come from those in the vicinity – in the research group, center, institute, laboratory, department or faculty where the research is conducted. The local peer review enables people directly in the environment to screen the work for its validity and integrity. Local peer review is not an adjunct to research, it is an integral part of the systematic investigation. Postgraduate research students need to present their work and findings to other research students and academics within the closest proximity to where the work was undertaken. If there are issues, then they can be best picked up by those familiar with the student; the facilities, equipment and instrumentation.

The next level of peer review – an independent level – should take place either through publication/conference presentation or by judiciously sending the work to peers/colleagues for assessment – and validation or repudiation. This provides an important step in the systematic investigation. Again, it is not an adjunct to the investigation, it is an intrinsic part of that investigation. The two levels of peer review should be considered as parts of the experimental process within the methodology.

Another key point to note in the context of a systematic investigation relates to the preparation of a thesis. This is often left by research students to the end of the research program – when it should be an ongoing part of the research documentation process, commencing at the beginning of the program. A thesis has a relatively rigid structure, regardless of:

- The field of research.
- The methodology.
- The experimental results derived from that research.

If the thesis is tackled as an ongoing part of the research program, then it can also serve as a systematic basis for the conduct of that research. For example, a thesis chapter on experimental design should be completed before a program of experimentation takes place – rather than as an afterthought to a collection of randomly conducted experiments. All too often, research students execute a research methodology without a rigorous template and, only during the late documentation of the thesis, discover that key elements were neglected or incorrectly performed.

Finally, research students need to discuss the context of their work in the broader field with their supervisor – in order to understand how one research project fits into a much broader field, and how significant the findings are within that field.

2.4 A Contribution to (Extension of) Knowledge in a Particular Field

Universities each have their own guidelines about the nature of the knowledge contributions that need to arise from postgraduate research programs. In addition to these, there are conventions within individual fields of study, and centers, institutes, departments and faculties may also provide more specific criteria.

From the supervisor's perspective, what constitutes a sufficiently large contribution to knowledge, to be accepted by external peers conducting an assessment, is subjective. It is further complicated by the nature of the research itself. If the research is very narrow in its scope, and a clearly identifiable gap in knowledge is filled – to sufficient depth – then that may satisfy requirements. If the research is multidisciplinary in nature, then clearly one cannot expect a research candidate to achieve the same depth of contribution in multiple fields – perhaps a broader definition is required.

Consider the diagram in Figure 2.1 – known as *Pasteur's Quadrant Diagram for Research (Stokes, 1997)*. This diagram quantizes research into various areas, based upon its applicability and its intrinsic knowledge content.

In determining what constitutes a reasonable contribution of knowledge from the research student, the supervisor needs to determine the region in which the student is conducting research. A student working in the *Bohr Quadrant*, say in a physics department, may need to close an identified gap in knowledge with a narrowband study, conducted to a substantial depth. A student working in the *Edison Quadrant*, say in an engineering department, may be seeking to take a range of existing knowledge components and apply them to a gap that exists in some application. The key points here are that

- The supervisor needs to understand the nature of the research outcomes and contributions.
- The supervisor needs to ensure that the research student understands the nature of the outcomes and contributions.
- The research student needs to be able to articulate, in his/her thesis and defense, the nature of the contributions to peer reviewers.

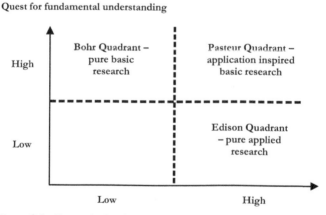

Figure 2.1 Pasteur's Quadrant Diagram for Research (Stokes, 1997).

Unless all these three criteria are fulfilled, then it is entirely plausible that those examining the research work will be out of synchrony with the research student and the project – and this could have serious consequences. If, for example, a research program is conducted in the *Edison Quadrant*, and examiners are expecting a high level of investigative depth in one narrow field, then they may undervalue the contributions of the student.

Deciding where the specific contributions of a postgraduate research program fit may be too onerous a task for the research student to undertake in isolation – particularly in the early phases of the research program. The supervisor's judgment, mentoring and negotiation with the student will be important in ensuring that there is some insight as to what will ultimately be required. In reaching this point, it may also be necessary for the supervisor to review his/her own assessment of the situation after taking advisement from others – particularly colleagues or peers in the field.

The starting point for these discussions between the supervisor and the student clearly needs to be the student's literature review, and the supporting case that he/she can make as to the need for the research, and the void that it fills – as determined by other scholars in the field. The next stage is getting a meaningful perspective, from published literature, as to what potential assessors may see as a sensible outcome.

The important issue here is that the research student and the supervisor need to acquire an intimate knowledge of the research examiners – not necessarily by name, but by nature and predisposition. And, while the research may be intrinsically significant in its own right, the reality of postgraduate study is that, ultimately, decisions of its worth and significance are made by an independent, external target audience. It is the supervisor's role to understand that target audience and to assist the research student in coming to terms with it.

2.5 Development of Rigorous Documentation Techniques and Disciplined Writing Style

2.5.1 Introduction

A postgraduate research student, having completed primary and secondary education – as well as at least one tertiary qualification – should be expected to have a reasonable ability to write and document a series of professional events and outcomes. However, the practical reality for many research students is that this is not the case.

If a research student is able to achieve great things in research but is unable to communicate them to peers in a systematic and compelling form, then what outcome has really been achieved by the postgraduate research study? Research writing is an integral part of the research process.

Importantly, research writing is something which needs to take place on an ongoing basis – a research student needs to document/record work accurately, on a day-to-day basis. This means not only keeping a good laboratory or note book but ensuring that meetings are documented, and the consensus agreed upon, to track progress. Documentation, and the organization of documents, becomes ever more critical the more busy the student's program becomes. So, instilling a *document/record everything* culture goes some way to creating a disciplined writing style. The thesis, progress reports and journal articles then become a logical extension of an everyday process.

There is a strong case to be made that it is the supervisor's responsibility to ensure that a research student develops this skill as a specific outcome of the research training program.

It is important to note at this point that one is not talking about having a research student developing a skill as a novelist. The writing outcome may be rather pedestrian in a literary sense. However, one needs to develop in the research student an accurate, logical flow of information. Prose, narrative and plot may have a role in some fields, but these are generally extras. A well-written research document may be dry but not tedious, because it should flow naturally. And, by flow, one also necessarily includes the information which needs to transfer from one brain to another – specifically, that of the reader.

Research supervisors often feel that it is not their role to teach students the art of research documentation and disciplined writing – some feel that it is beneath their dignity. However, the impartation of documentation techniques to the student is not an option. A supervisor cannot supervise practically unless he/she receives from the student properly written documents. Verbal reports do not pass muster because, ultimately, research knowledge needs to be transferred in written form. The research supervisor is the student's first reader/consumer of record. If the person closest to the student's research cannot read/understand his/her reports then what hope has anyone farther afield? In a practical sense, therefore, it is non-negotiable that the supervisor *owns* this problem and the responsibilities that go with it.

Universities generally have supporting departments which are able to assist students with preparation of theses and research papers – but these tend to be generalist departments, with no field-specific knowledge of the requirements of a particular area of research. The harsh realities of modern research supervision, and the nature of research students – particularly in an international learning environment – are that the research supervisor needs to take over-arching responsibility for student research documentation skills.

There are three aspects to writing for (and communicating) research. These are:

- A clear, succinct understanding of what the research is about – establishing a central research theme.
- An ability to present a rigorously accurate and compelling depiction of reality in the context of the conducted research program.
- An ability to write and structure documents in a concise and systematic form.

The starting point for all communication is that the research student needs to understand what it is that he/she is endeavoring to communicate. This may appear to be self-evident, but the lack of a clear understanding of the research project and its implications is a major underlying cause of communications problems, particularly in research dissertations.

It is particularly important that a research student is able to create historical documents (research papers and dissertation) that are based upon facts/evidence, or the carefully weighed opinions of learned scholars. Research documents also need to provide appropriate attribution/citation of the work of others, adopting whatever referencing scheme is accepted in the field of research. When there are a range of conflicting views in the field of discourse, the research student needs to be able to present these in a balanced and impartial manner.

In the context of the research activity itself, the research student needs to be able to present:

- A hypothesis.
- Supporting and contradictory arguments.
- Formal evidence/results.
- Critical self-evaluation of the research.

Overstatement of the significance of the research and/or outcomes needs to be avoided in order to present an accurate picture of reality. In particular, the value of the research needs to be placed into the context of the contributions of others.

All these things should already be familiar to an academic charged with research supervision but rarely are they apparent to a novice research student, who may be eager to please or eager to succeed – and thereby tempted to bypass/ignore a few facts, datasets and contrary learned views along the way.

The final aspect of developing a disciplined writing style pertains to the mechanics/semantics of the language being adopted. This presents considerable challenges for supervisors, particularly when students are not writing in their native language.

Herein, we examine some of the issues that need to be addressed in order to achieve the sorts of outcomes expected from postgraduate scholars.

2.5.2 Starting with Basic Research Objectives – A Central Research Theme

The starting point for research communication is that the author needs to have a clear understanding of the subject that is to be communicated. The unfortunate reality is that research students often do not understand what it is that they are attempting to communicate. This lack of understanding, combined with poor writing ability and complex technical material, can lead to disastrous results in a lengthy dissertation.

Research students can sometimes naively feel that if they express their research in simple terms, they are undermining its value. In fact, the opposite is generally the case, because the purpose of written communication is to build a bridge between the author and the reader. A complex starting point makes it difficult to build that bridge. Consider the Venn Diagram in Figure 2.2.

The research student author and a reader may have a significant amount in common in terms of the field of research. However, they are also two different people, each of whom has developed his/her perspectives of the world independently. In writing documents, research students need to understand that it is their responsibility to begin the communication process with a common area of understanding, in order to transfer knowledge to the reader. The common area of understanding is simple speech – free of technicalities and research jargon.

The challenge then is for the research student to be able to understand his/her area of research to the extent where it can be expressed succinctly, in terms that even a lay-person should be able to understand.

The problem with this exercise is well known. In the 17th Century, Blaise Pascal wrote *(Quoteinvestigator.com, 2012)*,

> *"I have made this longer than usual because I have not had time to make it shorter."*

2.5 Development of Rigorous Documentation Techniques

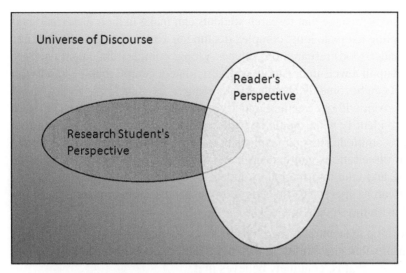

Figure 2.2 Understanding the basic problem of research writing.

The implications of the quote, claimed by numerous authors (including Mark Twain) since that time, are abundantly clear – that is, it takes time to create clear, concise communications. The problem is that research students can get themselves so busy *doing* that they don't take the important time required to *understand* and to *communicate*.

In order to achieve good communications skills outcomes, a supervisor should work with his/her research student until such time that the student can enunciate the objectives of his/her research in a few simple sentences. Until Pascal's writing objective is met, students have little chance of being able to communicate their research effectively – in oral, visual or written form.

It may be that a good approach to developing a compelling central research theme is to initially have the research student treat the exercise as an oral or visual task rather than a written one. How would the student communicate his/her work to an audience who are outsiders to the field? What would the opening slide of a short, succinct presentation look like?

A good analogy to the central research theme is what, in business circles, is referred to as an *elevator pitch*, wherein a person is required to sell an idea or product in the context of a short elevator ride with a potential buyer or financier. What is the most compelling description that can be made during the elevator ride? In postgraduate terms, this should be the central research theme.

A serious mistake that research students can make in their dissertations is in attempting to create long, complex documents based upon convoluted (i.e., poorly understood) research objectives. A long, complex document needs to be based upon a well-understood, well-articulated, central theme – otherwise the result can become incoherent for the reader.

Moreover, without a central underlying theme, it is difficult for a research student to identify what should be included in a complex document and what should be omitted. The end results can then be disappointing. For example, research dissertations can end up being submitted with irrelevant material included, and vital information excluded.

A research supervisor's objective early in the research program is therefore to work with the student to:

- Develop the central research theme.
- Ensure that the student understands and accepts the central research theme – that is, genuinely believes in it.
- Encourage the student to keep a printed declaration of the central research theme on display wherever he/she works.

This should prevent students from losing track of what they are supposed to be doing, and what they will need to communicate to others.

In the context of developing a central research theme, consider Examples 2.1 and 2.2.

Example 2.1 – Pretentious, Technocratic Summation of Research Objectives:

> *"The investigation of heterarchical, societal control structures (HSCSs) has been widely documented in a number of colloquial, quasi-complementary tomes in the sphere of sociology which draw succinct but peripheral analogies from the intrinsic behavioral patterns..."*

Example 2.2 – Thoughtfully Considered Summation of Research Objectives:

> *"The objective of this research was to answer a single question. That is, what are the factors that influence the controlling structures in a society?"*

The writing in Example 2.1 almost immediately alienates the reader, who is left with the task of decoding technical expressions and acronyms in order to get a glimpse into the world of the research student. The writing in Example 2.2

invites the reader into the researcher's world, with simple grammatical expression that naturally covers a common understanding between the writer and reader.

Importantly, the central research theme becomes the reference point for all the research student's subsequent activities. Specifically:

- Reviewed literature
- Experiments/surveys/tests
- Results and analysis

all need to be checked against the simple theme for relevance.

The central theme also sets the agenda for the research documentation process, including published papers and the final dissertation/thesis. Every sentence that goes into a research paper or dissertation is assessed against the central theme – if the sentence relates directly or indirectly to the theme then it is included – if it does not relate to the theme, then it may need to be excluded.

Research students can, and will, argue that it is not possible to express the objectives of their research in simple, clear speech – especially in complex areas of science or engineering. This is when supervisors need to step in and work with their students to ensure that they understand the broader, basic objectives of the research, devoid of specific technicalities and jargon.

The development of the central research theme is arguably the first and most fundamental outcome of any postgraduate research program – because all the other outcomes pivot around it. It maintains focus and clarity of purpose, and it prevents unnecessary mission creep in research objectives.

The central research theme will be examined further in the context of the preparation of a postgraduate thesis, in Section 12.5.

2.5.3 Developing Writing Discipline Early in the Research Program

It is important that the basic discipline of research writing is established early in a postgraduate research program in order to be effective, and to avoid a compounding of problems later. Documentation also needs to become an everyday activity for research students as a matter of systematic recording of events – it should not become an adjunct or an afterthought to the process which is only given consideration when a research paper or thesis needs to be written.

A logical starting point is for the supervisor to formally assess the student's initial capabilities in writing, as well as to introduce and instill the discipline of the writing style specific to the research field.

A particularly useful task to achieve these ends is to have the postgraduate researcher formally document his/her thesis introductory chapter or initial literature review, for assessment by the supervisor.

The emphasis here is in having the student present a complete and professional piece of work – not a *draft*, which can be a euphemism for unfinished/unprofessional work.

Supervisors need to develop the mental toughness to reject draft documents as a satisfactory initial outcome – simply because these tend to become the path of least resistance for both the research student and the supervisor. It is at this early, critical point in the research program that supervisors need to step up – with maximum effort – to ensure that a student's writing style becomes professional in a research context. This will provide major, ongoing benefits for both the supervisor and the research student throughout the course of the program – and ultimately time savings.

Disciplined writing is not only a basic requirement for research in its own right, but it also trains a research student to collect and structure his/her thoughts systematically for the purposes of research conduct.

Once a research student is ready to submit an initial document, and before the supervisor accepts it, a worthwhile exercise is to have the student go through and *pre-screen* his/her own initial submission – by parsing the work, sentence by sentence. A supervisor, based on his/her own experience in the specific field of interest, should be able to prepare a checklist for the student to use in parsing this submission. An example of a six-point checklist is shown in Table 2.4. This should serve to illustrate to the student that:

- Each and every sentence in disciplined research documentation needs to be watertight in its own right – that is, factually correct and independently verifiable.
- Research is about evidence, facts and data, and not about personal opinions.

Other specific elements to research writing that need to manifest themselves early in the research program include:

- Citation styles and consistency.
- An ability to present arguments in a cogent, systematic sequence.

- An ability to formulate forward research directions based upon a review of others' work, and based upon knowledge gaps identified by peers in the field.
- An ability to draw balanced and fair conclusions from the work of others.

A very common – but altogether naive and costly – mistake that a novice supervisor can make is to assume that a research student's writing style will fix itself through natural osmosis, and therefore improve along the way. Generally, it will not. Somebody has to do the hard work of ensuring that learning the art of research writing starts as soon as the research program does – and that somebody needs to be the research supervisor. Unchecked problems early in the program will exacerbate as the research student moves forward with the investigative process, and the remedial task will become more complex and multi-dimensional.

It is important here to reiterate that instilling writing discipline is a critical and integral part of the research learning process. For this reason, when research supervisors sub-contract this important teaching role to others, they

Table 2.4 An example of a sentence by sentence parsing tool to check writing discipline

Check	Assessment Criterion	Action
1	Is the sentence based upon the student's personal opinions?	Delete the sentence
2	Is the sentence factual – as supported by independently verifiable data or facts?	Delete if not factual – otherwise present/refer to verifiable data/facts
3	Is the sentence based on the opinions of other published scholars in the field?	Ok – include – but also include dissenting views for balance
4	If the sentence contains the opinions of other published scholars, and there are contradictory, published opinions elsewhere, does the sentence juxtapose/balance both arguments?	Ensure that varying scholarly opinions are balanced and fairly represented
5	Is there any unsubstantiated numerical phrasing/wording in the sentence? For example, does the sentence use expressions such as *"most"*, *"none"*, *"all"*, *"always"* or *"the majority"* without being able to factually justify the numerical implications?	In the absence of other factual substantiation, replace numerical expressions with non-numerical equivalents (e.g., *"many"*, *"few"*)
6	Does the sentence contain *lazy phrasing*? For example, *"...people have often stated that..."*	Name the people or quote supporting data

are potentially undermining the value of disciplined research writing in their own field.

A research student also needs to understand that the discipline of research writing isn't merely about creating neat research documents. It is about developing the underlying, rigorous, systematic thought processes that bring about that disciplined writing. This is something which can only be taught effectively by the supervisor in the context of the specific field of research and the writing conventions therein.

2.5.4 The Issue of Grammar and Writing Competence

An important aspect of developing a disciplined research writing style pertains to the intricacies of language and grammar. The world's *Top 10* universities, as classified by the Academic Ranking of World Universities *(Shanghairanking.com, 2015a)*, are all English speaking and so is their published research. It is therefore difficult to get around the fact that, for most research students – regardless of their originating countries – English will be the required language for their dissertation and research publications.

This clearly presents a problem for those who have English as a second language and, in most English-speaking universities, there is a significant number of research students for whom this is an issue.

There are two internationally accepted English language testing regimes that are in common use in universities around the world. These are:

- TOEFL (Test of English as a Foreign Language).
- IELTS (International English Language Testing System).

The TOEFL system has a range of individual scores (from 0–30) for reading, listening, speaking and writing *(Ets.org, 2015)*. The IELTS has a range of overall performance bands that cover reading, listening, speaking and writing *(Ielts.org, 2015)*. The IELTS bands range from 0 to 9, with a Band 9 performance indicating complete proficiency with the language, and a Band 1 performance indicating no real ability to use English as a language.

Universities tend to use TOEFL and IELTS scores to determine entry prerequisites for students wishing to undertake postgraduate research programs. In some cases, supervisors will never encounter those who do not achieve appropriate proficiency ratings because they will have been screened out by university administration during the postgraduate application process. However, supervisors will eventually need to come to terms with how well these two scoring systems reflect the requirements for postgraduate research students in their field. And this will generally only come about after extensive personal experience with students who have achieved varying levels of proficiency.

It is not uncommon for students to perform reasonably well in either TOEFL or IELTS and still have considerable difficulties with developing a research writing style. After all, students with English as a second language have the dual tasks of coming to terms with the complexities, nuances and inconsistencies of the English language, as well as the complexities of a disciplined research writing style.

At some point, regardless of the TOEFL/IELTS scores that a research student has achieved prior to admission to a postgraduate research program, the supervisor will need to determine whether or not the student is capable of independently generating a complex research document in the form of a research paper or dissertation. In some cases, the answer will be no – and this causes considerable dilemmas for the student, the supervisor and the university.

The basic notion of a postgraduate dissertation is that what is presented to examiners for assessment is the actual work of the student. But what exactly is being assessed? Consider the following issues as examples:

- Are students in physics, chemistry or mathematics to be deemed incapable of high level research in their field because their English grammar and phrasing are poor?
- Do students conducting research into mathematics need to have the same level of English proficiency as those undertaking research into English literature?
- What happens if an engineering student is capable of writing concise, systematic arguments in a well structured sequence – only to be let down by shortcomings in phrasing or the nuances of language?

There are two opposing schools of thought on these sorts of issues. Many universities believe there is a case for providing editing support to research students who have English as a second language, because they are ultimately being assessed in terms of their competence in their field of research, and not on their English grammar skills. There is also a case to be made for not providing any editorial support for research students, on the grounds that examiners should be reading a document prepared entirely by the student – with all its intrinsic grammatical errors and inconsistencies – rather than one which has had expert editing elsewhere – and which does not reflect the writing capabilities of the research student.

There is a possible middle ground on this issue. And that is an acceptance that, for those students who are undertaking research into areas where language and grammar are not the pivotal/central parts of the assessment, writing support should be provided to create a professional dissertation. This provides a well written historical document for future research reference. The assumption of the examiners is then that the quality of the writing is ignored in the assessment, on the grounds that it may or may not be the work of the research student. The downside of this approach is that future employers may assume that the graduating student has writing skills commensurate with the dissertation – and to this extent the thesis may prove to be an inaccurate reflection of reality.

Notwithstanding the problems that may arise in future careers, one also has to consider that technology has largely overtaken the argument about the originality of grammar and phrasing. After all, it is possible to use common word processing features to correct or improve spelling and grammar – and isn't this the same as the university providing professional human support for the exercise?

Accepting that a significant proportion of research students will require (and get) language support for the preparation of their dissertation and research papers, the issue that needs to be resolved is who will provide this?

Research supervisors generally don't see it as their role to do editorial work for their research students, but there are efficiencies to making such an effort. Firstly, the generalist editorial support service provided by universities can sometimes create as many problems as it resolves. Even common words and phrases can have a different meaning when used in the context of a specific field of research. So, it is possible that generalist editorial staff can confuse commonplace words – which have been chosen to convey specific information in a research context – as being incorrect in a grammatical context. The end result can be a completely muddled document which appears to be grammatically correct but is incorrect in the context of the research field.

Secondly, the writing style of a generalist editorial support unit can be out of touch with the style in which particular research fields are documented. This means that the supervisor has to undo work which has been already been changed by the editorial unit.

Thirdly, there is an inefficiency in having a research student submit work to an editorial support unit at the university, only to then have the supervisor make further changes, and create an iterative chain of editing bureaucracy.

If, on the other hand, a supervisor makes the effort of performing both the technical and grammatical editing of a first chapter (e.g., the literature review or thesis introduction), then the student is likely to fall in with the supervisor's preferred approach, and the subsequent editing load and editorial changes diminish significantly.

Finally, the supervisor's efforts in editing his/her research student's work have collateral benefits – specifically, the supervisor can maintain a close oversight of the student, and make ongoing assessments of capabilities, limitations and deficiencies that need to be addressed. Progress towards the end objective can be monitored closely, and the supervisor is always acutely aware of which work has emanated from the student. This isn't always possible with third-party editing.

2.6 Development of Oral/Visual Research Presentation Skills

The ability to communicate the outcomes of research in an effective and compelling manner to a live audience is an important outcome of the research training process. Oral and visual presentation skills can also be particularly useful in focusing a student's mind on clear, succinct arguments that can subsequently be used in research papers and theses.

All research students, at some point in their programs, will be required to present their work to an audience – either at in-house seminars within their own research group/department/faculty, or externally at professional conferences.

The common problems with research student presentations include the following:

- An inability to précis the research into a straightforward introduction.
- An inability to construct a systematic sequence of information morsels which convey the research in a concise and intelligible manner.
- Lack of perspicuity – no insight into what the audience wants to hear about the program – a disconnect between the speaker and the audience.
- An eagerness to present everything the research student knows about the subject, rather than key points and findings.
- A temptation to provide overly long presentations that lose the audience interest.
- General oral communications problems – poor diction; soft-spoken presentation voice, mumbling, constant use of time-wasting words – *"ums and errs"*.
- Poor visual presentation skills – visual slides that are too complicated and contain too much information; slide text too small for audience viewing; too many slides; transition speed between slides too fast, etc.

Some of these problems can be resolved by basic improvements in technique but many presentation problems arise because students do not have a clear picture in their own minds as to the nature of their own research or its contributions. In Section 2.5.2, the importance of the central research theme was emphasized. In oral/visual communications the central theme is also critically important.

When a research student begins his/her presentation to an audience, the first task is to make a connection with that audience and, as in writing, this comes through the introduction of the research theme, in simple terms, intelligible to the lay-person.

From a supervisory perspective, it may be necessary to work actively with a student in developing presentations, until they become comfortable with the expectations. A good approach – particularly when supervising numerous students – may be to develop a presentation template that ensures that research students present in a concise, systematic and disciplined manner. Table 2.5 provides an example of a presentation template.

Even within the context of a structured presentation template, research students are likely to want to present everything they have learned about a particular area, rather than key findings. A case of too much information and not enough knowledge transfer. One approach to stopping this is for the supervisor to impose strict time constraints that force students to decide what is relevant or irrelevant to their research theme.

In some countries, there are national programs which encourage postgraduate research students to compete with succinct, sharp presentations by creating a short-time-presentation environment. One such program is the *Three Minute Thesis* program *(Threeminutethesis.org, 2015)* – which, as the name

Table 2.5 Example oral/visual presentation template

Slide	Content
1	Basic Identification – Institution, Department, Researcher Name, Title of Presentation
2	Central Theme of Research
3	Background Issues/History (Who? What? Where? When? Why?)
4	Summary of Specific Work Being Presented – Experiments, Designs, Surveys, etc.
5	Summary of Results Derived
6	Comparison/Benchmarking with Earlier Research
7	Broader Implications of Research Findings
8	Conclusions

suggests, asks students to present their research theses – theme, outcomes, etc. – within a three minute timeframe. In that program, only a single presentation slide is permitted – without animations or transitions. Points are deducted for presentations which are rushed. The objective is to create a disciplined approach to creating a compelling oral/visual presentation of research.

Another approach to developing these skills is the five/ten approach – where students present for five minutes but need to defend their work by answering questions from peers for a further ten minutes after their presentation. This Socratic approach encourages research students to think long and hard about what is presented and how well they have been able to communicate their message.

In the absence of a specific local program, there is nothing to preclude research supervisors from working together to develop their own in-house activities to build presentation discipline among research students in their own research group, center, department or faculty. Technology can also provide useful supporting devices. For example, *(DeSantis, 2012)* and *(A Blog Around The Clock, 2016)* look at the advantages and disadvantages of using networking tools, such as Twitter, to encourage students to come up with very short summaries of their work. The objective is to use writing brevity as a vehicle for focusing the thought processes behind the research – and to ensure that the research student is absolutely clear on what he/she is endeavoring to achieve.

Regardless of the training method that is adopted, the basic objective/outcome is to establish a mindset which seeks to rank presentation quality (i.e., knowledge transfer) above presentation length. In so doing, this should encourage the research student to critically evaluate his/her own research directions and findings.

2.7 Peer Evaluation/Acceptance of Research

Jonathon Swift, the author of Gulliver's Travels, once remarked,

> *"That was excellently observed', say I, when I read a passage in an author, where his opinion agrees with mine. When we differ, there I pronounce him to be mistaken."*

And therein lie the strengths and weaknesses of peer review. They are often based upon the peculiarities and prejudices of humans. Notwithstanding their limitations, in research, peer reviews still represent the best means of assessing the validity and limitations of research. Like many other aspects of research, peer evaluation isn't an annoying adjunct to the research, it is an intrinsic part of the professional process.

Postgraduate research students are subjected to a range of different peer reviews during the course of their program. The most obvious one is that of the research supervisor, who should be providing peer review on an ongoing basis. Of course, supervisors should never be the sole peer reviewers because they advise students on research directions and can therefore never be entirely independent.

For this reason, supervisors need to encourage their research students to subject their work to as many outside peers as they can. Some of the peer review mechanisms will be formal – such as in refereed journal publications – but various informal mechanisms are also valuable.

The informal peer review mechanisms that are available to the research student are numerous and include:

- Review amongst fellow research students – group meetings are a good way of getting local peer review as well as practicing presentations in a supportive environment and without pressure from academic staff.
- Consultation with other academic/research staff in the research group, center, department or faculty.
- Consultation with academic/research staff in other local universities.
- Electronic correspondence with academic/research staff at an international level.
- Online *preprint* on recognized scholarly websites that are designed to elicit immediate feedback from peers.

The objective is to ensure that during the course of the research program, and prior to final examination/defense, the researcher learns about the:

- Range of peer perspectives on his/her research.
- Strength with which various views are held, and the predominance of any particular views among peers.
- Strengths, limitations and/or potential fundamental flaws in his/her research methodology.

Ideally, these views should also be ascertained prior to submission of research findings to refereed journals for formal assessment.

The outcome is not just an understanding of the level of acceptance of the research, but also a determination of the sorts of views that are held by potential future examiners. This will assist the student in formulating a Socratic defense for his/her work, with a full awareness of the arguments that need to be made in its support.

Importantly, peer review assists research students in understanding that a research dissertation is not a book, wherein an author's views can be expressed to a general audience as unsubstantiated truths. A dissertation is aimed at a very specific, targeted audience – and the research student author needs to write accordingly.

It is also common in various research fields for differing, conflicting views to be held firmly. In the context of the final assessment, whether or not the student's research ultimately represents a universal truth is then a matter for the judgment of the target audience – based upon the Socratic defense mounted by the research student.

There are other outcomes which arise from the ongoing use of peer review within the confines of the research unit in which the postgraduate student is stationed. In particular, local peer review provides the highest level of scrutiny of the research. Within the confines of a university research group, it is likely that everyone knows what everyone else is doing – so, when conflicts arise about who contributed what elements to a piece of research, they can be resolved in-house. Additionally, within the local research group, research staff and students are generally acutely aware of whether various experiments were actually conducted – and, therefore, the chances of individuals fabricating data can be minimized.

In some circumstances, informal peer review is difficult to achieve – particularly when a research student is part of a larger collaborative program, and intellectual property (IP) and confidentiality issues come to the fore. However, this is one aspect that needs to be carefully considered when research contracts are negotiated – because removing internal peer review mechanisms is effectively removing an integral part of the research student's learning paradigm.

2.8 Published Research Findings

The publication of postgraduate research findings is one of the most obvious outcomes of the learning program, and it also constitutes an important mechanism for formal, external review of the research. Additionally, the formal publication of research work effectively stakes a claim, on the part of the research student, to a contribution of knowledge. It may also be beneficial in a career sense as an achievement in its own right.

The specific benefits include:

- A formal peer acceptance of the research student's research approach and findings.

- Exposure of the research work to other scholars in the field.
- A formal contribution to the research output metrics of the university and of the supervisor who may be a co-author of any published work.

There are some caveats that need to be considered when supervisors are seeking to publish work. These include:

- Is the work being published as a significant research outcome in its own right, or because of pressure from the university to publish work?
- Is the work that is being published a complete and accurate depiction of reality – or just a partial depiction of reality?

These are critically important issues. Publications need to arise for the right reasons – that is, as genuine contributions to knowledge in the field – and not solely as a career benefit for the supervisor or student.

The second question – the complete depiction of reality – is one that every supervisor needs to address. Does the publication only represent positive outcomes of the research program, or does it reflect the negatives as well?

If the research program has been structured correctly in the first instance – with an *open hypothesis* – wherein all possible results are a useful contribution to knowledge, then the issue of a complete depiction of reality is a moot point. However, if the research program has been initiated with a *closed hypothesis* – such as,

> *"The objective of the work is to demonstrate the efficacy of an increased dosage of Xylamine in patients with..."*

then there are serious issues that need to be addressed. Specifically, would the research still be published even if the results were negative for the closed hypothesis? If not, then why not? Ultimately, if the objective of research is to give a complete picture of reality, then both positive and negative results need to be published.

In some areas of research, particularly in medicine/pharmaceuticals, this is a serious issue, because a failure to publish negative outcomes can lead to invalid conclusions from the broader community. For example, if ten trials are conducted on the efficacy of a particular pharmaceutical, and nine negative results are not published, then publication of the tenth (positive) gives a statistically incorrect picture of the field – even though the tenth study may be statistically valid in its own right.

In recent years, there has been growing concern in the research community about the publication of predominantly positive results, without balancing publication of the negatives. In the medical research field, various groups have banded together in organizations such as AllTrials *(AllTrials, 2015)* to lobby for registration and publication of both positive and negative research findings. More broadly, in a world increasingly placing information online, the publication of an incomplete picture of research is likely to be exposed in any event, and potentially damage the reputation of the research student, the supervisor and the institution.

It is the responsibility of the supervisor to ensure that, firstly, research is conducted in a genuinely open manner – to an *open hypothesis* – that does not predispose outcomes into a particular form – or render some results worthless because they prove the hypothesis negative. Secondly, the supervisor needs to utilize his/her experience in ensuring that publications arising from the research are genuinely valid and meaningful in the broader context of the field in which they are published – and in consideration of all the available evidence.

Another important consideration in terms of publication of work is how many publications a research student should produce during the course of his/her studentship. It is imperative that this decision is driven by the academic merit of the outcomes arising from the program, rather than the supervisor's personal career imperatives. Finally, research supervisors should discourage students from *gaming* the research metrics system, by dividing what should be a single research paper into Part A, Part B, etc. in order to maximize research metrics.

2.9 An Understanding of the Broader Perspective of the Research and Research Methods

Research students generally need to conduct their research within a narrow band of interest, and to a reasonable depth. For this reason, they can end up viewing their own research through the aperture of a *straw* – with a very narrow interpretation of its significance.

The supervisor's role is to work with the student to develop a greater understanding and appreciation of the research and its outcomes. These include:

- The relative contributions of the research within the broader field, and within a historical context.
- The significance of the research methodology within the current field, and potentially other fields.
- The potential commercial applications of the work and its value to various products or services.

The research student needs to have a good understanding of the relative importance of his/her work. If the research student's work is a three-year study, in the context of thousands of person-years of effort in the same field, then this needs to be observed within the final dissertation, so that the importance of the outcomes is not overstated (exaggerated).

On the positive side, there may be genuinely useful outcomes arising from the research that go well beyond the results/findings themselves. In some instances, research results may lead to the development of improvements in commercial products or services, and some consideration needs to be given to the relative contribution of the research to such commercial outcomes.

2.10 Establishment of a High Caliber Research Team in Which the Research Student Is an Active Member

A novice academic/researcher may commence supervisory activities with a single research student, but rarely will that student be working in complete isolation. Initially, the supervisor and the student may both be part of an existing research group, center, department, institute or faculty. As the supervisor develops and assumes an increasingly senior role, he/she may work towards having a new research group with a team of research students.

The transition from supervision of a single student to supervision of a research group may take place before the first student has even completed his/her research program. For this reason, the supervisor needs to work towards building a team – in which the research student will be an integral part. Traditionally, the entire structure of academia has centered around individualism and individual excellence. However, given that many research students will not ultimately work in an academic environment, it is hardly reasonable to limit their future professional development by creating loners who work with a silo mentality.

Academics seeking personal excellence generally find it difficult to work in a team, and because the environment tends to create competition and rivalry, it is not generally a good model of teamwork. Sometimes, the university model is better suited to the creation of mavericks than institutional builders and team players. The modern research supervisor, however, needs to work towards making the academic model more reflective of contemporary practices in the outside world. Unless he/she has had management training on team building then there is a case to be made for background reading on the subject to ensure that the research student is working in a healthy, functional and productive environment.

Ultimately, in order to be successful in knowledge advancement, a research supervisor needs to:
- Attract high caliber individuals as research students.
- Retain them as postdoctoral researchers.
- Ultimately, collaborate with them as academic peers.

Invariably, high-caliber individuals in any field are both mobile and desirable, so it is all the more difficult to attract and retain them if a research group is dysfunctional. A good research supervisor should work towards ensuring that his/her research group is functional and positive.

2.11 Creation and Maintenance of a Dynamic Research and Learning Environment Which Actively Encourages the Exchange of Ideas and Opinions

The creation of a functional research group, in which individual members feel comfortable, is not sufficient to attract and retain high caliber postgraduate researchers. There are a number of key factors that need to be addressed in order for a research grouping, perhaps led by the supervisor, to become a magnet for talented individuals. These include provision of:

(i) *Quality Resources* – technical, administrative, information technology (IT) and office space. An institution that wants to attract world-class research students needs to be world-class in basic infrastructure. Not all institutions can be world class but, at a minimum, resources need to match institutional research goals.

(ii) *Quality Staff* – smart students want to mix with other smart students and smart academic staff.

(iii) *Challenging Research Programs* – high caliber research students want to work on high caliber projects that can have significant transformative outcomes at a global level.

(iv) *An Inclusive Environment* – where the research students are welcome to provide opinions and suggestions – and where their intelligence is valued in practice not just theory.

(v) *A Fun Place to Work and to Be* – intelligent people require more than just intellectual stimulation in an academic sense – they need to be able to engage their sense of humor as well.

(vi) *Connectivity* – research is global, and local opportunities in any given field are limited, so research students need to feel that their environment

is a respected node in the global research effort in their field. Connectivity should also extend to relevant business and industry.
(vii) *Ongoing Research Funding* – research students need to feel that there is some potential to stay on and have a career in their research group.

A novice supervisor is unlikely to be able to address or provide all these sorts of facilities in his/her own right. However, an acute awareness of the need for them is critical to steering any existing research groupings towards these goals.

2.12 Establishment of a Research Team in Which Peer Review of Each Member's Work Is an Intrinsic Part of the Research Effort

In Section 2.7, it was noted that peer review – which takes place as close as possible to the conduct of the research – is the most effective form of evaluation and feedback. Within the confines of the research group, center, department, institute or faculty, it is possible to provide close scrutiny over the research that has been performed, and to have a reasonable level of certainty that the person claiming credit for the work is the one who actually performed it.

Once research leaves the confines of the intimate research grouping – and goes to external publication – then the opportunities for effective assessment are diminished. It is not realistic to expect referees working for journals to be able to determine whether research results presented to them are genuine or have been embellished – or even completely fabricated. The principles behind publication in international journals are all based on a level of trust in the university in which the research is conducted. There is an implicit assumption that the university has integrity measures in place to ensure that what is presented for publication consideration is:

- Research which has genuinely been undertaken by the authors of the research paper.
- Based on genuine evidence as described in the research paper – and not on fabricated data/information.
- Original content that has not simply been reproduced from some other source without appropriate attribution.

As a starting point, within the university environment and as they pertain to postgraduate research, these checks are the responsibility of the postgraduate supervisor. Further, there is an onus on the supervisor to inculcate a culture of peer review within his/her own team as an intrinsic part of the research process. Apart from the conduct of in-house research seminars, there needs to be an understanding within the research team that hypotheses, methodologies, results and conclusions need to be evaluated by more than just those conducting the research.

Formal, bureaucratic methods often beget formal, bureaucratic responses. So, if a supervisor can achieve a culture where his/her research students willingly and enthusiastically offer up their research for ongoing assessment by their fellow research students, then a significant outcome has been achieved in the context of postgraduate research integrity.

2.13 A Dissertation/Thesis for Examination and Defense

2.13.1 Nomenclature

In this book, the terms thesis and dissertation are used interchangeably – as they are in many universities around the world. In some North American institutions, the thesis often refers to research work presented for a Master's qualification, and a dissertation refers to research work presented for a Doctoral qualification. Herein, given the lack of global uniformity in definition, there are no distinctions made in terms of nomenclature based upon postgraduate qualification levels.

2.13.2 The Supervisor's Role in Thesis Preparation

An academic charged with the responsibility of postgraduate research supervision will already be well aware of the challenges associated with the development of a thesis; its examination and defense. However, each novice research supervisor will have generally had only a singular experience with the process – that is, his/her own thesis – and so there is something to be gained by looking at the broader perspective.

The thesis is the principal tool for assessment of postgraduate research but it is also:

- A historical document, catalogued for reference by future researchers.
- A handbook on how a hypothesis is explored via a particular methodology and investigative process.
- A document which brings together scholarly literature from relevant fields in a comprehensive review that backgrounds the research.

A common error amongst postgraduate students is to view the thesis as an afterthought to a program of investigation, rather than a document which develops and drives the investigative process in a systematic manner. It is altogether common to find theses which have been written after the research event, and which contain too much irrelevant material and not enough of the material that is required to substantiate the hypothesis and investigative process.

An important point which many novice researcher supervisors and research students fail to appreciate at the outset of the program is the generic structure of a thesis. In general, a sound, basic thesis structure can be independent of the field of research and independent of the hypothesis and investigative process. Consider Table 2.6 as a generic example of a thesis structure.

Regardless of the specific chapter breakdowns that are chosen, the basic attributes of the thesis, as exemplified in Table 2.6, do not change greatly, whether the research is in the field of business, economics, physics, social science, engineering or medicine.

Once a research hypothesis is identified – and expressed in the form of a central research theme – and an initial literature review conducted, then there

Table 2.6 Example of a thesis structure

Chapter	Title	Content
1	Introduction	Presentation of the Who, What, When, Where and Why basics of the research program and overview of the hypothesis and approach
2	Literature Review	A review of scholarly literature that places the current research into context and highlights gaps/deficiencies identified by other scholars that the current research is intended to fulfill
3	Methodology	An explanation of how the hypothesis is to be tested and evaluated
4	Experimental/ Investigative Design	A guide which documents the systematic investigation of the hypothesis through experimentation or evaluation of other independently verifiable information
5	Results of Investigative Process	A presentation of the results of the investigative process
6	Broad Context Discussion of Results	A discussion on the broader implications of the results and a critical self-evaluation of the identified shortcomings in the current research program
7	Conclusions and Recommendations	Based upon the results, benefits and shortcomings of the current research, recommendations for future investigations.

is no reason why the bulk of a thesis cannot be written up until the point of the results and discussion sections. The key point here is that the thesis preparation process should start at the beginning of the postgraduate research program – and not at the end.

It is not uncommon to come across research students enthusiastically performing experiments and gathering data, only to find at the end of the process – when they do attempt to reverse-engineer a thesis around what they have done – that the hypothesis, methodology and results do not come together as a whole. The research supervisor's task is to avoid this scenario by having research students rigorously document their methodology and investigative process in a thesis that develops alongside – and often preceding – the experimental/investigative process. To a large extent then, the thesis is a research outcome that should emerge before much of the other work.

2.14 A Professional Career Foundation

The area of investigation for a postgraduate research student tends to be very narrow in scope – often intentionally so, in the expectation that it will be explored to significant depth. However, by the same token, this means that career opportunities for those emanating from postgraduate research programs are equally limited – in the context of the research field itself.

Modern postgraduate research programs came into existence in the 19th Century, when the first research Doctorates were awarded. In the early years of postgraduate research, only a handful of elite students were chosen, from each final-year class, to move from a basic degree to an advanced research degree. This small cohort was chosen to be given an apprenticeship in research so that it could move on to fill faculty positions within the university. In the 21st Century, circumstances have changed considerably and, yet, postgraduate research programs are still largely based upon a research apprenticeship model, which focuses upon the creation of university academics and researchers.

The reality of the modern world is that the number of people emanating from universities each year, with postgraduate research qualifications, is an order of magnitude larger than the annually available number of tenured positions within the global university system. Of course there are also many non-tenured positions available within the university system, and there are also commercial and government research and development facilities where research degree qualifications can be of value. Nevertheless, the practicality that research supervisors need to face is that the bulk of their research students may not ultimately work within the university sector. Many will not even work in any field of research.

If it is to be at all meaningful in the context of a career in the modern world, the postgraduate research learning program needs to address these hard facts. To do less than this is to do a disservice to the research students.

For these reasons, the research investigation which is undertaken during postgraduate study is now seldom the underpinning element of the research student's long-term career. However, if the research training during the program has been conducted professionally then there are numerous other attributes that can service a broad range of career options. Specifically, these relate to a postgraduate researcher's ability to:

- Ascertain the current state of knowledge in a given field through systematic exploration of published scholarly work.
- Design and undertake a rigorous and disciplined investigation.
- Act in such a manner that personal biases and antipathies are removed, and to balance arguments from a range of different perspectives.
- Work independently to perform systematic analysis of problems, information and datasets and to determine their significance.
- Write in a disciplined, systematic manner, devoid of personal opinions.
- Develop recommendations based upon hard evidence and without personal bias.

These should be the crux of a postgraduate research program and form the basis of a long-term career in leading-edge organizations – whether they be in the academic arena or the commercial/entrepreneurial or government sector.

In addition to these attributes, another output of a postgraduate research program is the high level of tenacity which is engendered, and which is required to persevere through a lengthy investigative process.

All these traits are generically applicable ones which are important in commercial, as well as academic and government research and development.

The research supervisor can, however, also improve upon these basic attributes by ensuring that the research student has sufficient supplemental training (either formal or informal) in areas such as:

- Entrepreneurship.
- Venture capital.
- Product/service commercialization.
- Team building.
- Financial management.

In some universities there are specific programs which encourage and support research students to use their research work as the basis for start-up companies or as tools for other commercial ventures.

At the core of it all, the supervisor has an important role to ensure that the research student is fully aware of all the skills and attributes that he/she has acquired during the course of the research program, and how these can best be put to use in a range of possible career/business choices.

A career foundation cannot, however, be built within a few weeks after a research dissertation is submitted for examination. The process needs to start early in the research program. The supervisor needs to determine:

- What future aspirations the research student has.
- Whether the student's personal attributes (i.e., strengths and limitations) are applicable to the desired career/business choices.
- How adjustments/supplements to the basic research program can be used to help achieve these ends and to help address any underlying limitations.

Importantly, the research student needs to be given an honest and practical assessment of how feasible his/her career aspirations are, particularly if they are in the university sector. It is not uncommon for supervisors to exaggerate the prospects of a long-term career in academia when, statistically, the chances of these being fulfilled are limited. Importantly, if academic prospects in a particular research field are limited, then one of the student's background tasks during the research program will be to determine ways and means of establishing a career outside academia – perhaps in commercial research; more generalized professional practice, or management.

Some research students may also seek to commercialize their research work and start their own companies. A research supervisor needs to understand how and when this will take place – and whether it fits within the intellectual property arrangements that are in place for the postgraduate project.

2.15 Underpinning Research Outcomes for Future Research Grants

A postgraduate research program has a limited time-span but, for the academic supervisor, there may be ongoing work and research requirements within the institution. Before embarking on a postgraduate research supervision, the research supervisor needs to understand the broader picture for his/her own professional research activities, and how the postgraduate program fits into these.

A postgraduate research program, which is an end in itself, may have only limited value, but one which can be used as the basis of future competitive research funding can have multiple benefits. Firstly, if the competitive research funding provides resources for postdoctoral positions then it may be possible to retain the research student after graduation. Secondly, additional funding may enable the supervisor to expand the scope of the current postgraduate research in order to achieve more significant outcomes in the field.

At this point it should be noted that it is generally not a research student's role to actively pursue funding, but a supervisor should have it in his/her mind that the postgraduate research investigation may be a good basis on which to base future research grants.

The timeline for many competitive research funding schemes can range from several months through to an entire year – from initial submission to receipt of funding. For these reasons, consideration needs to be given to research grant applications as soon as a postgraduate research program starts yielding meaningful results.

In some instances, the research undertaken by postgraduate students may be applied in nature, and the pathway may not be a traditional research grant but, rather, precompetitive research and development, which may require commercial/seed/venture capital funding. If this is the case, then the research supervisor needs to consider the option of packaging early postgraduate research outcomes into an attractive value proposition, which can then be presented to potential commercial sponsors. Some of these may wish to form a partnership to take the research further at the conclusion of the program – others may wish to invest in the intellectual property or, perhaps, enter into a royalty arrangement.

2.16 Research Outputs that Feed into University Research Performance Metrics

Despite the altruistic motives that universities or academics may have, the reality is that modern institutions work in a nationally and globally competitive environment. In order to secure research funding, institutions need to demonstrate performance in the form of accepted metrics, which include the following:

- Research publications.
- Citations for publications.
- Patents.
- Awards (e.g., Nobel Prize, Fields Medal).
- Competitive research grant income.
- Endowments.

- Doctoral research student completions.
- International collaborations/joint publications.

An intrinsic limitation of all metrics, and all parametrically-driven performance systems, is that humans tend to *game* the system in order to maximize numbers for personal self-interest or advancement. Sometimes this occurs at the expense of institutional building, and other times at the expense of colleagues or, worse still, the basic truths which researchers have a duty to uncover and protect. It is important supervisors understand the performance environment in which they operate but, more importantly, that undesirable *gaming traits* are not passed on to postgraduate research students.

Research student activities will clearly contribute to some or all of the performance metrics, depending on the nature of the research; the capabilities of the research students, and their productivity. The decisions relating to "how much?" a research student contributes are left to the subjective determinations of each research supervisor, so it is imperative that the supervisor acts in the best interests of:

- The research student.
- The integrity of knowledge and learning.

To do less than this, and to use research students as tools for personal career advancement is not only unethical, but it is an activity that can metastasize through an entire research grouping or department – ultimately destroying any institution-building capacity, as well as damaging the credibility of the unit, and potentially making it dysfunctional.

2.17 Establishment of a Long-Term Professional Relationship between the Graduated Student and the Supervisor

The Reverend Russell Conwell, founder of the Temple University in the United States, recited a famous speech over 5,000 times between 1900 and 1925. The speech was entitled *Acres of Diamonds (Americanrhetoric.com, 2015)*. Contained within that speech was the story of a farmer, who had considerable wealth and a successful farm. After a priest told the farmer of the value of diamonds, and the places where they could be found, the farmer became restless and discontent with his lot in life. He sold his farm and abandoned his family in order to search for the enormous riches of diamonds, only to ultimately perish in his attempt. The man who purchased his little farm subsequently found that that farm ironically had one of the world's greatest riches of diamonds just below the surface:

> *"Thus...was discovered the diamond-mine of Golconda, the most magnificent diamond-mine in all the history of mankind, excelling the Kimberly itself. The Kohinoor, and the Orloff of the crown jewels of England and Russia, the largest on earth, came from that mine."*

The moral of the story is readily apparent – and that is to not go searching elsewhere for riches that may already exist in one's own lot.

Universities expend considerable time, energy and resources cultivating relationships with external bodies in government, business, industry and alumni. The objective is to seek research funding or endowments that can be used for the betterment and advancement of the institution. However, each of those people in government, business and industry was once a student – and all are humans who can remember – fondly or otherwise – their school and university lives.

It is important that every supervisor understands that every research student can become a potential benefactor to the university and to research in the future. Further, there is little point looking for external benefactors when many potential benefactors exist in one's own *farm*. The end of the research supervision process should not become the end of the relationship between the supervisor and research student graduate – it should be the beginning of a new professional relationship. It is the supervisor's role to cultivate that relationship by maintaining contact with the research student as they move onwards and upwards through their career. If the supervisor has performed his/her role fully, ethically and fairly during the period of research candidature, then this is something that the graduating research student should embrace.

3

Basic Responsibilities of Research Supervision

3.1 Overview

It has already been noted herein that the supervision of postgraduate research students encompasses far more than just the conduct of systematic programs of investigation. It is ultimately about the development (personal and professional) of human beings, as well as their safety and welfare.

In any institution, there will be numerous rules, procedures and guidelines that need to be considered in relation to supervision, and these include:

- National/regional laws.
- Institutional rules/procedures.
- Faculty/Departmental rules/procedures.
- National/international conventions or codes of conduct for research in particular fields.

The consequences for breaching the formal rules – particularly as they relate to the health, safety and welfare of an individual – or abrogating managerial responsibility in their enforcement, can be severe. In some jurisdictions, at national/regional levels, these can include sanctions such as fines and imprisonment or, at institutional level, termination of employment.

The relationship between a research supervisor and a research student is therefore not just a casual, professional relationship, but one in which the supervisor needs to assume a range of particular responsibilities – often the same as those that would apply when managing a regular employee of the university.

The specific areas which are covered in the subsequent sections of this chapter are:

- Health/safety.
- Personal welfare of the research student.

- Confidentiality of the research student and research.
- Protection of intellectual property.
- Research ethics.
- Prevention of discrimination, harassment, bullying.
- Personal/professional ethics development.
- Personal development.
- Selection of research examiners.

3.2 Health/Safety

The most basic requirement for any research supervisor is to ensure the health and safety of the research student. In many jurisdictions, this requirement is not just a moral or institutional issue but one with serious legal ramifications and sanctions.

The World Health Organization (WHO) defines Occupational Safety and Health in the following terms *(WHO.int, 2015a)*:

> *"Occupational health deals with all aspects of health and safety in the workplace and has a strong focus on primary prevention of hazards. The health of the workers has several determinants, including risk factors at the workplace leading to cancers, accidents, musculoskeletal diseases, respiratory diseases, hearing loss, circulatory diseases, stress related disorders and communicable diseases and others."*

Further, the WHO defines work related diseases in the following terms *(WHO.int, 2015b)*:

> *"An 'occupational disease' is any disease contracted primarily as a result of an exposure to risk factors arising from work activity. 'Work-related diseases' have multiple causes, where factors in the work environment may play a role, together with other risk factors, in the development of such diseases."*

For each country, and often regions within that country, there are specific legislative requirements for maintaining the health and safety of employees. Each university, operating under these jurisdictions, therefore creates its own internal policies and procedures in line with local legislative requirements.

Notwithstanding any specific local legislation and university regulations/procedures, it is likely that a research student will need to be treated in the same way as an employee in terms of health and safety. The research supervisor may therefore have significant legal and procedural responsibilities in addition to the self-evident moral ones that pertain to health and safety in supervision.

The levels of risk for the student are clearly dependent on many factors, including:

- The field of research.
- The nature of any practical experimentation.
- Exposure to (usage of) equipment/machinery, devices, radiation, chemicals, biological materials/agents, etc.
- General university laboratory and office environment.

A research student conducting research in an area such as business or economics will have vastly different health and safety issues to one working in areas such as bioscience, chemistry, physics or engineering. However, in all cases, the supervisor needs to make some assessment of the potential risks to the student operating in his/her environment – this assessment should be over and above any formal processes that the university already has in place.

For low level risks, an informal mental assessment may suffice but as the levels of risk associated with the environment increase, so too does the formality of the risk assessment and mitigation. For example, a student working solely in an office environment, using a computer, may only require basic consideration of seating/ergonomic arrangements associated with the computer workstation. A student working in a chemistry or bioscience laboratory may need to undergo a complete occupational health and safety training program in order to function safely in that environment.

In addition to basic risks associated with the research student's broad program of experimentation, there may be other risks – unforeseen at the beginning of the research program. For example, an engineering research student may spend the bulk of their time doing analysis and modeling on a computer but, at some point, may need to cut metal samples on an electric saw for experimentation purposes.

The key point about health and safety is that the supervisor cannot make assumptions about what a research student knows or doesn't know about the environment and risks. It is the supervisor's role to determine the risks and formally ensure that the student is aware of them.

In most universities, there are a range of procedures and regulations relating to health and safety of staff and students. These include elements that are:

- Legislated under national/regional law.
- Global to the entire university (e.g., ergonomic seating posture for use of computers).
- Specific to a faculty.
- Specific to a department.
- Specific to a research center.

Additionally, even within a small research grouping, there may be supplemental rules and procedures that a research student needs to understand. For example, in an Electrical Engineering department, an electrical power laboratory group may stipulate that no individual is permitted to work on high power circuits without the presence of another person trained in Cardio-Pulmonary-Resuscitation (CPR). Within the same department, people working in another research group with low-power battery-driven devices may not require the same level of safety scrutiny and training.

In some cases, educational institutions are given special exemptions from various national/regional laws in relation to the operations of their laboratories. This is because, sometimes, generic legislation would completely preclude any experimentation. For example, legislation may state that workers other than electricians must not be exposed to any *live* electrical terminals or wires – this may be relaxed in the case of an electrical laboratory training students.

In addition to procedures and regulations, any well-operated university should have in place training and induction programs that facilitate execution of those procedures/regulations – particularly for research students. A research supervisor needs to be aware of which training and induction programs will pertain to his/her research students as they operate in their specific environments of research.

In the absence of any formal training/induction programs, or specific procedures/regulations, a research supervisor still needs to take a moral and proactive responsibility for the health and safety of research students. The creation of a detailed risk assessment and risk control form for every conceivable type of activity is an onerous task, often undertaken by specialists in the field of health and safety. However, as a bare minimum, from a supervisory perspective, a practical starting point can be to create a simple checklist for each research student – considering risks and mitigating strategies as they pertain to the specific field of research. An example is shown in Table 3.1.

In addition to physical health and safety issues, a supervisor also needs to consider the issue of stress. Specifically, is the supervisor's behavior, or

Table 3.1 Developing a health/safety risk assessment checklist for research students

Activity	Risks to Health/Safety	Preventative Actions/Training/Induction
Computer work	Posture/Back injury	Ergonomic workstation
Use of milling machine	Cuts, entanglement, eye injury, amputation	Formal training/oversight by laboratory technician; First-aid/safety training
Transfer of test samples to furnace	Back strain/injury, burns, eye injury from sparks	Training on furnace use; training on lifting procedures

the behavior of other staff/students in the area likely to cause a postgraduate student stress? This requires careful consideration because many small things which may, on the surface, appear inconsequential to the supervisor might create serious emotional issues for the student. For example, consider the following situations:

- An introverted, international research student is placed in a office environment full of boisterous, local research students.
- A timid research student with English as a second language is asked to present research in a public gathering of staff.
- A research student who, unknown to the supervisor, is working long hours; has additional family responsibilities, or has to do outside paid work to make ends meet, is told that he is not working hard enough.

These are all things which can cause enormous stress for a student – and potentially related health problems as well. These issues are covered further in Section 3.3.

The key point that needs to be taken from all these discussions is that the novice research supervisor cannot assume that the issues of health and safety relating to the research student are automatically covered by the university. It is often the supervisor's role to formally interface the research student to university health and safety procedures. In some cases, the supervisor may also need to formally intervene to ensure that the interests of the research student are fully considered.

Prior to allowing any research student to commence his/her studies, it may be useful for a supervisor to consider the following course of action:

(i) Ensure personal familiarity with university regulations and procedures as they pertain to health and safety of staff and students.
(ii) Determine the training and induction programs available within the university for various aspects of health and safety.

(iii) Identify faculty, departmental, center, institute, laboratory or research grouping related safety procedures and induction programs.
(iv) In cognizance of the suite of training and induction programs at relevant levels, ensure that the research student participates in relevant activities.
(v) Develop a simple risk assessment/mitigation template for each student as an *aide de memoire* and a possible action plan for the specific research program.
(vi) Consider the individuals, the environment and the supervisor in the context of the stress that they may impose upon a research student – the supervisor should put himself/herself in the research student's shoes to try and identify potential sources of stress and to avoid subjecting the student to them.

In addition to all these factors, a research supervisor also needs to consider the health and safety of research students who specifically need to work outside the university – for example,

- In an affiliated university research facility (e.g., teaching hospital).
- At the premises of a commercial partner organization.
- In an external setting while conducting surveys.

In these cases, a research student may need to adhere to more than one set of regulations and procedures – that is, those of the university and those of the external environment provider.

3.3 Personal Welfare of the Research Student

In addition to the physical health and safety of a research student, a research supervisor – as a mentor acting in a one-to-one professional relationship – also *de facto* assumes some degree of responsibility for the personal welfare of the student. The level of responsibility obviously depends upon:

- The environment.
- The specific nature of the research student.
- The level of moral responsibility that a supervisor wishes to assume.

The specific areas of personal welfare that need to be considered include:

- Psychological welfare (stress, emotional wellbeing, etc.).
- Personal issues (living arrangements, financial problems, relationship and family problems, etc.).
- Environmental issues (research environment and colleagues).

Clearly, some of these areas will be outside the purview of the supervisor and, in any case, the supervisor may have no capacity to resolve the underlying causes. Importantly, most supervisors are not trained to deal with, or provide, psychological counseling to students, and this is a task that should be left to experts. However, a supervisor may need to be astute enough to recognize problems and symptoms in order to refer a student for expert support before a crisis occurs.

Universities in general have a broad range of support mechanisms and staff to assist students with personal problems. The difficulty is often getting the student to acknowledge that he/she has a problem in the first instance – especially given that the supervisor is unlikely to be an expert in even assessing this. It is often also the case that the supervisor is the root cause of a student's problems or angst, and is therefore blind to the problems that exist.

If a supervisor suspects that a research student is experiencing personal problems – or has indeed been approached for support by the student experiencing them, then there are some basic things that can be done as a starting point. The following list may be helpful, based upon the conduct of informal discussions with the student:

- Without being too demanding or prescriptive, see if it is possible to get the student to elucidate on his/her problems.
- Determine whether the student is able to identify specific causes of the problems – for example, an argument with other staff or students.
- Determine whether the student is able to put forward his/her own preferred solutions to the problem – and best possible resolution mechanisms.
- If the resolution mechanisms are within the jurisdiction of the supervisor, and are reasonable and readily achievable, then enact resolution measures – otherwise, ask the student for time to consult with others and organize another meeting with the student.
- Consult with colleagues in the research group, center, institute, faculty – as appropriate. If the response is inadequate, contact the appropriate university level support staff associated with student counseling and seek guidance.
- If university level support/counseling staff see the matter as urgent, then contact the student as soon as practical to organize meetings with counseling experts.

In the final analysis, despite the best intentions and efforts of a supervisor, a research student may refuse to seek formal help from professional university support/counseling staff. A supervisor has no practical mechanism to coerce a student to do so. However, depending upon the relationship that the supervisor and student have with other members inside the institution, it may be possible to broach the matter judiciously with the student's colleagues in order to elicit more informal support – keeping in mind the strict requirement not to breach the student's confidentiality.

The seriousness of personal problems must not be underestimated and, self-evidently, if these become seemingly insurmountable from the perspective of the student then a supervisor needs to face the reality that there may be a risk of suicide. Each university should have its own processes/procedures for dealing with the issue of possible suicide, however the University of Cambridge (UK) Counseling Service describes well why a student may attempt to commit suicide in the following terms, using a list of ten possible causes *(Counselling.cam.ac.uk, 2015)*:

"For some students suicide will follow a period of depression while for others it is likely to be an impulsive act, perhaps triggered by a traumatic experience, for example the death of a loved one, or by a relatively insignificant event which may be seen as the 'final straw'. The following is a list of some of the feelings and experiences that may contribute to someone feeling suicidal":

1. *Loneliness – developing into the all-consuming feeling that there is no-one there and that no-one really cares or will notice whether they live or die. The suicidal person can feel totally alone and isolated.*
2. *Feelings of hopelessness and helplessness – where the student feels that no matter what s/he does, nothing seems to get better, and that no-one is able to help.*
3. *Feelings of worthlessness, of being 'a waste of space' – s/he will never amount to anything, and that any care, interest, or encouragement shown would be unjustified or based on a false premise. Convinced they are not worth caring about, such people are likely to have very low self-esteem, and to not readily accept compliments or praise.*
4. *Depression – in those who are clinically depressed, their perceptions of themselves, others and their situation are usually unduly negative. Many of the above feelings are not only common but are also felt to be unquestionably true. Although observers may be clearly aware of the depressed person's talents, achievements, and of others who care deeply about them, it is their internal perceptions that need to be taken into account.*

3.3 Personal Welfare of the Research Student

5. *Plans falling through – especially where the goals have considerable personal importance – e.g., not settling well at university, the break-up of an important relationship, or a student not achieving her/his academic goals. As a result the student can feel inadequate, a failure, ashamed, unlovable...*
6. *Inappropriately high levels of stress – of the kind experienced by those with exceptionally or unrealistically high personal or academic expectations. Students can easily come to feel stressed by academic demands and for some there will be times when the level of stress becomes unbearable. Those who have been high achievers, in particular, can feel that their academic success is crucial to their personal identity, and if the former is under perceived or real threat, their identity is also endangered. To such people, the idea of not getting a 1st can be felt as an utter and unbearable humiliation. The Oxford study... on student suicide quoted above found that of those who had committed suicide "two-thirds of the students had been worried about academic achievement or their courses". However, the Collins study... at Cambridge did not find an increase in suicides around examination times.*
7. *Anger – suicide can sometimes be seen as the act of someone who is very angry, perhaps even as an act of revenge, (for example after the ending of a relationship).*
8. *Alcohol and drugs – for some a suicide attempt may be an impulsive act when under the influence of alcohol or drugs. In this state a person may seriously underestimate the risks of their actions, and be more vulnerable to the above feelings.*
9. *A history of mental or physical illness.*
10. *Feeling overwhelmed – when problems in a number of areas of life occur at the same time – for example academic problems, a family crisis, and the ending of a relationship – the sense of pain may be overwhelming."*

Note especially how the factor listed under Paragraph 6 pertains specifically to issues and personality traits of postgraduate research students.

A supervisor should make himself/herself aware of the possible warning signs – either from the list above, or from more specific information at his/her institution, to ensure that crises and tragedies do not arise because of a lack of information or action within the institution.

It is good institutional practice for universities to ensure that students have staff other than the supervisor to whom they can turn in the event of problems – particularly because it is often the case that the supervisor is – either knowingly or unknowingly – the root cause of the problems. Some institutions insist on a minimum of two supervisors in order to provide pressure relief – others have supervisory oversight committees. If these sorts of mechanisms are not available, then it would be beneficial for a supervisor to create his/her own by, perhaps, providing either an unofficial second supervisor or mentor/advocate for the student.

3.4 Confidentiality of the Research Student and Research

A research supervisor needs to be able to deal with a range of confidentiality issues, specifically those that relate to:
- Personal discussions with the research student.
- Assessments and appraisals of the research student's work.
- Departmental/faculty/university level discussions and correspondence in relation to the research student.
- Intellectual property (IP) arising as a result of the research.
- Negotiations with external collaborators.

In cases that relate to personal details/discussions/correspondence with or about the research student, the notion of confidentiality will generally be based upon the subjective, professional judgment of the supervisor – sometimes supplemented with university procedures or conventions.

The obvious question may then be,

"How does one develop subjective, professional judgment?"

The answer to this question is in fact relatively straightforward and that is, when in doubt, as a starting point, always err on the side of confidentiality. As one develops supervisory and management skills, one also develops a greater capacity to determine what is or is not appropriate in terms of confidentiality. Often, one will develop a greater degree of latitude as one matures.

A good rule of thumb is to assume that everything which is spoken or written in the context of professional duties relating to the student should pass the *public domain test*. That is, anything which would be deemed unacceptable if it was to reach *the public domain* should neither be articulated nor written in the context of private discussions or correspondence.

In cases related to IP and negotiations with external parties, the mechanisms for dealing with confidentiality are normally set out in research contracts and are generally straightforward. However, supervisors need to get advice on how these can impact on the conduct of a postgraduate research program. For example, an IP agreement can sign away the rights of a supervisor and research student to publish research papers – or even to publish the thesis – or even to allow external examiners to assess the thesis. Before signing any confidentiality arrangements in relation to IP or external research collaborations, a research supervisor needs to take advice from university legal representatives.

In cases where a research student is required to sign a confidentiality agreement, as a result of participating in a collaborative research program which is the subject of a legal agreement, the research supervisor has an obligation to advise the student to take his/her own professional advice on the matter before signing or agreeing to anything.

It is important for the supervisor to explain to the student that, in negotiating any collaborative agreements, the university's legal representatives act to protect the university's interests – not necessarily those of the research student. The research student may therefore need to take external counsel before signing any documents pertaining to an agreement between the university and an external partner.

In between the hard copy contractual law confidentiality and the judgment-based confidentiality of professional relationships there are also intermediary issues. For example, a research supervisor may have it in mind to commercialize research work emanating from a postgraduate research project. The research student performing that project may have it in mind to publish as much as possible about the work, with the intention of moving on to another area or career, regardless of the commercial potential of the research.

In these instances, it is the research supervisor's responsibility to discuss possible research commercialization plans with the research student during the formative stages of the project. Both parties need to agree on the pathway that will be taken and, if there is an agreement on project commercialization, then any confidentiality arrangements may need to be determined – preferably in writing or by formal contract – at the outset of the program to avoid future disputation.

3.5 Protection of Intellectual Property (IP)

A postgraduate research supervisor may be required to act on behalf of the university in negotiating and protecting IP arising from a project undertaken by a research student. There are a range of possible scenarios, including:

- The postgraduate research project is part of a larger program of research activities for which the university is developing a suite of IP.
- The research student is working in a collaborative research program which has been generated by the university with an external partner.
- The research student is a full-time employee of a company and has elected to use the research and development being undertaken in the company as the basis of a postgraduate research degree within the university.
- The research supervisor can see long-term commercial potential in the postgraduate research outcomes, and wishes to secure IP for possible commercialization.

In all these cases, the protection of IP only has significant merit if the researchers and the institution have the resources and willpower to:

- Monitor usage of IP.
- Litigate against unauthorized breaches.

In many instances, the cost of litigation can far outweigh the value of the IP and hence the benefits of protection are a moot point.

In order to understand the benefits and limitations of IP protection in the context of a university, one has to have some insight into the role of research within the commercialization process. The often quoted *rule of thumb* for commercialization is the so-called *1:10:100 Ratio*. That is, for every dollar expended on research, ten dollars need to be expended on development and a hundred on commercialization. In some areas, such as pharmaceuticals, the ratio may be considerably higher.

In the best case scenario, therefore, a university research outcome will generally merit less than one per cent of net commercialized income – and that is assuming that the research contributes to the entire commercialized product or service. In most cases, university research may only contribute to a small part of a larger product or suite of services, and so the potential income will be accordingly smaller.

The key issue that the research supervisor needs to consider prior to expending university resources on the protection of IP is whether there is enough potential income to even mitigate the cost of protection, much less make a net profit.

There may be alternatives to attempting to gain directly from the IP arising from research – that is, more efficacious mechanisms than the traditional patent protection and royalties. One approach may be to consider that there are benefits from collaborative research with industry and, in a research partnership, it may be worthwhile signing over IP in exchange for ongoing collaborations and research scholarship funding, rather than royalties. These may form a tangible fixed payment, with a degree of certainty, rather than the an undefined probability of income from royalties at some unknown later date.

In situations where the research student is the principal originator of the research, it may also be worthwhile to consider allowing (and supporting) the research student to commercialize the research independently. In such a situation, the university may receive no short-term financial benefit but – if a start-up company becomes financially successful – and the graduated student is sufficiently grateful, then he/she may give back to the university in far larger value than simple royalties.

All of these possibilities need to be weighed up and possibly discussed with the university's legal and IP departments before selecting a forward pathway. Once a direction is chosen, the supervisor needs to determine how this will impact on a research student's research program and future career. Specifically,

- Will the research student be permitted to publish during the program and submit a thesis to external examiners?
- Will the final thesis be permitted to be available as open-access in libraries and online?
- Will the research student be permitted to use the work in his/her future career – or start his/her own business based on the research?

Clearly, all these questions have serious implications for the research student and his/her future career. At some point, therefore, in order to make any university decision meaningful, the research student will need to agree to certain conditions/restrictions.

Importantly, it also needs to be noted that in some countries, government scholarships for research students may be conditional on research students making the findings of their research public – and perhaps not signing away IP. These sorts of snares need to be identified at the outset of the program.

Only when all these factors have been resolved can a final decision be made on the IP arrangements for a postgraduate research project.

3.6 Research Ethics

There are two types of ethics for which a supervisor has to assume responsibility in postgraduate research. The first is the formal set of rules which are in place in universities, hospitals and other research organizations to govern experimentation pertaining to humans and animals. The second is the general, moral set of human operating principles by which the supervisor and student need to live their professional lives. In this section, we consider the former set of ethics issues.

Each university should have its own set of governing procedures pertaining to the conduct of research which has some impact upon living beings – either human or animal. In larger universities these will be governed by a specific ethics department – in smaller institutions they may be included within the activities of a broader research governance structure.

Research supervisors, who have themselves conducted research in human/animal experimentation, will already be familiar with the onerous requirements to carefully document and substantiate proposals for conduct of research procedures in these areas – and to have these formally analyzed and approved by a university body responsible for ethics. Sometimes, however, the need for ethics proposals and approvals may come as an unexpected requirement for a supervisor undertaking work in a seemingly innocuous area. For example:

- A researcher wishing to undertake an online survey in social science, involving human participants, may be required to submit the process to an ethics committee for approval.
- A researcher wishing to gather personal information that is openly available on social media, use it for analysis, and then publish the results which disclose individual identities may be required to get ethics approval.

For those who are new to the process of seeking ethics approval, there are a few elements that need to be considered:

- The process can be onerous.
- The process can take a considerable length of time to complete.
- Ethics approvals may be conditional on researchers undergoing formal training programs and gaining certification in order to conduct particular types of research.

In the United States, at Harvard University, for example, researchers undertaking research relating to humans – including surveys – require formal certification, undertaken by bodies such as the Collaborative Institutional Training Initiative (CITI) of the University of Miami. CITI operates a range of ethics based certification courses including

- Animal Care and Use *(Citiprogram.org, 2015a).*
- Human Subjects Research *(Citiprogram.org, 2015b).*

Other institutions around the world may have in-house or generic training certification programs that also need to be completed before approval can be given for particular types of research.

In terms of research supervision, because of the length of time associated with the process, the issue of ethics approval requires detailed consideration at the commencement of the program. A supervisor needs to discuss with his/her student whether any elements of the research will cover areas that require ethics approval and, if so, how the time associated with the process can be accommodated within the research plan.

The issue of ethics is further compounded when research students are operating in multiple research environments – for example, a student may conduct research experiments within the university proper and also at an affiliated teaching hospital. The teaching hospital may have different research ethics procedures and guidelines. It may therefore be the case that each set of experiments requires separate application and certification processes, further complicating the student's research plan.

The key points to take from these discussions are that:

- Supervisors cannot assume that they will be exempt from formal ethics requirements – they need to check this through university regulations at the outset of the program.
- Some seemingly innocuous human-related investigations (e.g., simple surveys) may require ethics approval.
- Ethics approvals and certifications are time-consuming and need to be incorporated into student research plans.
- Submissions for ethics approvals need to be carefully prepared and scrutinized by the supervisor – an ethics committee rejection of an ill-conceived proposal could delay a student's research for weeks or months.

3.7 Prevention of Discrimination, Harassment/Bullying

Most democratic nations have enacted a system of national/regional laws to cover the issue of discrimination in various forms, generally:

- Disability
- Ethnicity
- Gender
- Race
- Religion.

Universities within those nations naturally operate under the umbrella of those legislative requirements and may also impose their own additional procedures and sanctions as they pertain to the treatment of staff and students. In some countries there are also additional legal mechanisms in place to prevent the harassment and bullying of individuals.

A research supervisor who undertakes activities including:

- Recruitment of research students.
- Selection of research students.
- Awarding of scholarships/stipends to research students.
- Awarding of resources/facilities or conference travel stipends to research students.
- Management and assessment of research student activities – at a one-to-one level – during the research program,

has significant legal and procedural responsibilities in relation to prevention of discrimination, harassment and bullying. Importantly, the supervisor is not only responsible for his/her behavior but may also be responsible for the behavior of others in the environment in which he/she exerts control.

Each nation/region (and university) has its own specific definitions of the terms discrimination, harassment and bullying. Supervisors need to be aware of their local definitions. However, the following broad definitions, from *USLegal*, are particularly helpful to understanding the challenges that supervisors need to address.

The concept of *Discrimination* is defined in the following terms *(Definitions. uslegal.com, 2015a):*

> *"Discrimination refers to the treatment or consideration of, or making a distinction in favor of or against, a person or thing based on the group, class, or category to which that person or thing belongs rather than on individual merit. Discrimination can be the effect of some law or established practice that confers privileges on a certain class or denies privileges to a certain class because of race, age, sex, nationality, religion, or handicap."*

The concept of *Bullying* is defined in the following terms *(Definitions. uslegal.com, 2015b):*

> *"Bullying is generally defined as an intentional act that causes harm to others, and may involve verbal harassment, verbal or non-verbal threats, physical assault, stalking, or other methods of coercion*

such as manipulation, blackmail, or extortion. It is aggressive behavior that intends to hurt, threaten or frighten another person. An imbalance of power between the aggressor and the victim is often involved. Bullying occurs in a variety of contexts, such as schools, workplaces, political or military settings, and others."

The concept of *Harassment* is defined in the following terms *(Definitions. uslegal.com, 2015a):*

"Harassment ... is generally defined as a course of conduct which annoys, threatens, intimidates, alarms, or puts a person in fear of their safety. Harassment is unwanted, unwelcomed and uninvited behavior that demeans, threatens or offends the victim and results in a hostile environment for the victim. Harassing behavior may include, but is not limited to, epithets, derogatory comments or slurs and lewd propositions, assault, impeding or blocking movement, offensive touching or any physical interference with normal work or movement, and visual insults, such as derogatory posters or cartoons."

Some supervisors – particularly novice supervisors without significant management experience – believe that their newfound status as a supervisor gives them *carte blanche* to dictate to their research students what they should and should not do. The definitions above clearly demonstrate that this is *definitely not* the case, and modern management and legislative frameworks simply don't function on the *my way or the highway* dictum.

Much and all as academics believe that anything which does not lead to research outcomes is bureaucracy, the fact remains that that bureaucracy is generally backed up by legislation and university sanctions – up to and including termination of employment or even imprisonment.

Universities often have formal staff training and certification programs that cover the issues of discrimination, harassment and bullying – and the sanctions that apply to those who breach protocols. Supervisors therefore need to ensure that they:

- Are up to date with university practices and regulations in these areas.
- Consult with relevant university bodies before recruiting/selecting research students to ensure that their intended procedures comply with university policy.
- Always act in a manner which would be deemed – by an impartial observer – to be fair, reasonable and unthreatening.

3.8 Personal/Professional Ethics Development

A research student will often view the research supervisor as a mentor, and may ultimately end up adopting that supervisor's professional traits in later life – for better or worse. The supervisor therefore has a *de facto* responsibility to ensure that the traits passed on to the research student are for the better.

In addition to the formal ethics requirements and certification related to the specifics of the research program, the supervisor needs to consider how his/her behavior is subliminally forming professional attitudes in the student.

A supervisor may wish to consider how he/she would tackle the following issues/challenges of professional ethics insofar as they relate to management of postgraduate students – all of which have been derived from actual student complaints:

- Is the supervisor acting as a mentor or tormentor? Is the supervisor going to elicit good work practices and productivity by personal positive example, or by threats and intimidation?
- Will the supervisor pass on bad traits (bullying, overworking students) inherited from his/her own supervisor under the dictum that *it never did me any harm*?
- Is the supervisor capable of separating his/her own personal career requirements from the best interests of the student – *"writing this research proposal for me will be good experience for you"* – or *"you should publish this paper with me as first-named author because you will get more citations"*?
- Will the supervisor use the *not now I'm really busy* excuse to avoid his/her supervisory responsibilities?
- Will the supervisor ask a research student to dissect a single research project into multiple part-papers to increase his/her total publications and because *it will be good for the research student's research career*?
- Will the research supervisor ask his/her research students to all cite each others' papers to increase all their citations because *it's good for the research group to have more citations*?
- Will the research supervisor ask research students to include his/her name on all papers because he *contributed to all of them by virtue of getting the grants that paid for their scholarships*?
- Will the research supervisor ask that a research paper bear his/her name even though he/she has neither read nor contributed to the paper?

- Will the supervisor ask the student to delete/omit data points in their research which don't fit the supervisor's hypothesis because *they are extraneous outliers*?

The issues listed above constitute some of the more common complaints against supervisors, and there are a number of key points to take from these examples:

- Research students are generally highly intelligent and are capable of seeing through transparent self-interest in a supervisor who feigns that behavior as being in the best interests of the student.
- Research students will ultimately graduate – some may become highly successful – and they are unlikely to give back to an institution which has staff which they hold in contempt.
- Some research students will go on to become supervisors in their own right and perpetuate the sorts of problems of professional ethics illustrated above.

A research student needs to respect his/her supervisor. In turn, the supervisor needs to earn that respect – not just by being an expert in a given field but also by acting in a genuinely ethical manner. Respect will never be found on a curriculum vitae littered with dubious publications, citations and grants, regardless of how these appear in research metrics.

Finally, it should be noted that a lack of professional ethics is largely self-defeating for the supervisor. Most senior university staff are acutely aware of the sorts of behaviors outlined in the list above and, once they are detected, regardless of other achievements, an academic will find his/her own behavior counterproductive to moving upward in a career sense.

It is worth remembering that a reputation takes decades to build and it can be wiped out very quickly with opportunistic behavior.

3.9 Personal Development

The research supervisor and the research student both need to consider their participation in a postgraduate program as a journey of personal development. However, in the case of the supervisor, acting as a mentor, there is also an onus to look beyond career development self-interests and towards those of the research student.

It may be uncharitable to suggest that some supervisors view their research students as little more than a low-cost form of research labor that can be exploited for their own personal career development. If this is the case, then there is little that can be said to change such a mindset, save to say that such an attitude/approach has already been tried and tested by many academics – and found to be wanting in the long-term. In the end, truth will out, and naked opportunism is not a sustainable pathway to research eminence or respectability. It is also counterproductive to the extent that self-serving behavior is generally transparent to others. Colleagues who might otherwise be supportive, given the presence of goodwill, may in fact become aggressive obstacles to progression.

For those research supervisors who value the true essence of research and education, and wish to use their time as supervisor to create something of lasting value – in the form of a highly successful research graduate – then there are a number of things that need to be considered.

First and foremost, it is generally too late to look at careers for the postgraduate research student at the end of the research program – by that stage, either the student has made his/her own arrangements or has decided to leave the academic research environment – perhaps with some bitterness. The time to start the process of personal development is at the outset of the research program.

A good approach may be to develop a table of possible development pathways for one's students, with estimates of the requirements and timeframes involved. An example is provided in Table 3.2, but supervisors should consider developing their own sets of development pathways, based upon the research field/profession in which they operate. These can be used as the starting point for discussions with the research student early in the research program.

In looking at the sorts of issues and timeframes associated with career development for the research student, a few points become apparent:

- The research supervisor has to acknowledge/accept that the research student may not want to stay on as a postdoctoral researcher at the conclusion of the research program.
- If the research student is seeking a career in academia, and the supervisor believes they have the talent to make a significant contribution, then the supervisor needs to be able to present a convincing case for why the student should stay on.
- If a research student makes it clear that they wish to pursue non-academic career pathways, then the supervisor has a potential role in providing support in the student's development – perhaps funding relevant training programs from a research budget; providing networking opportunities for the student, etc.

Table 3.2 Example of possible development pathways and requirements

Career Aspirations	Key Factors	Timeframe
Academic Career	• Significant research outcomes • Publications/conference attendance • Citations • Academic service work – reviewing for journals, contributing to student programs • Networking with senior staff in current and prospective universities • Understanding of broader university research and education requirements and performance issues	3 Years
Professional Career in Broad Field of Research	• Understanding of research outcomes in the context of business/industry • Experience in/exposure to (e.g., internship) relevant companies • Understanding of company business models in fields of interest • Understanding of competitive advantages of PhD qualification relative to other graduates	1–2 Years
Management Career	• Formal management training/accreditation (e.g., MBA) • Basic business/industry experience (e.g., internship) in possible future employment organizations • High level communication ability • Networking with business-oriented colleagues	3 Years
Commercial Research and Development Career	• Demonstrable ability to deliver tangible research outcomes – on time • Ability to communicate R&D outcomes in an efficient, concise manner • Experience in commercial R&D environment (e.g., internship) • Possible Management training/accreditation (e.g., MBA)	2 Years
Start-up Company	• Understanding of basic entrepreneurship concepts and processes • Financials, budgeting, generation of business plans • Understanding of the business value proposition of the research outcomes • High level of communication ability • Networking with business angels, seed funding and venture capital organizations • Networking with enthusiastic colleagues to form possible teams	3 Years

If a research student gets to the end of their research program and starts filling out job applications in the hope of *finding something interesting*, then the supervisor has to recognize that they may not have fulfilled the broader personal development aspects of the mentoring role.

It may seem unfair that a research supervisor needs to extend himself/herself into the career development role but such is the reality of the modern world, where many who graduate from postgraduate degrees move outside the academic research arena.

In the final analysis, a research student who has performed a sound piece of research but is unemployed or unemployable at the end of the research program is as significant a failure for the supervisor as one who has performed poorly and failed the program outright. The research supervisor's role is to see that neither of these negative outcomes occur, and this requires considerable extra work on his/her part beyond the normal technicalities of research supervision.

3.10 Selection of Research Examiners

In the final phases of the postgraduate research program, a supervisor will need to have some involvement in the identification/selection of examiners for the student – for the thesis and/or other defense, as applicable.

The level of involvement varies from institution to institution. For example:
- Some universities allow the supervisor to explicitly nominate those who will conduct the examination.
- Some universities ask the supervisor to present a list of possible examiners (which are independently short-listed by others).
- Some institutions have sufficient resources to identify examiners completely independently of the supervisor.
- In some US universities, a thesis committee is formed after a student has passed qualifying exams – that committee ultimately formulates the nature of the examination process.

Whichever approach is applicable, the supervisor needs to be acutely aware of those who will conduct the research examination process – perhaps not by specific name but certainly by research predisposition.

It has already been noted herein that, in the context of postgraduate examination, a research thesis is not regarded as a book aimed at a general audience. While research findings may be generic, the thesis is written for a

3.10 Selection of Research Examiners 77

specific target audience. In addition to detailing the research investigation and findings, a research thesis also serves to mount a defense of the methodology; directions and decisions made during the program. A supervisor needs to have an insight into whether a student's defense is feasible given the target audience.

By way of example, consider that in a given field of research there may be two, or more, well-established schools of thought on the possible approaches. Those who represent one particular school of thought may reject outright another school of thought – regardless of the merits of the argument. In these circumstances, the supervisor needs to determine if it is fair to expect a student to mount a defense for one school of thought which, in all likelihood, will never be accepted, under examination, by those working in another.

In order to set the groundwork for a fair examination of the research student's work, a supervisor needs to:

- Ensure that he/she is aware of the varying schools of thought on the scholarly approaches associated with the research, and the strength with which convictions are held by peers in the field.
- Consider particular individuals that may not be appropriate for examination purposes because of their strongly held views (universities generally provide an avenue for exclusion of particular examiners because of their beliefs or views).
- Consider the audience for which the thesis has been written and the target and nature of the defense contained therein.

Once these issues have been considered, there is the task of nominating potential examiners. It is not uncommon for academic supervisors to consider nominating well-known colleagues to conduct examinations – however, this requires considerable reflection. Specifically,

- Is it really fair to a student, who has completed several years of work in a particular field, to have his/her work simply *rubber-stamped* by a sympathetic colleague?
- Conversely, is it fair to a student if a supervisor's colleague feels that they need to be tougher on that student's work just because of a close professional relationship with the supervisor?

There is merit to the case that the fairest approach to an independent, impartial examination is to engage an independent, impartial examiner who has no personal connections to either the supervisor or the student. This creates problems in its own right because it means that a supervisor has to use *cold-calling* techniques to contact academics and determine if they would be prepared to undertake the work required to assess a thesis. However, the proliferation of professional networking social-media sites means that it is possible for supervisors to have an ongoing cyber-connection to a large number of scholars in the field – and potential examiners can be sought from these contacts.

4
The Relationship between the Supervisor and the Student

4.1 The Role of the Supervisor

Among the common complaints, cited by students in their retrospectives on postgraduate research, are those that relate specifically to the supervision that was provided to them during the course of their study. Many of these complaints have arisen because of the misalignments between supervisor and student expectations during the course of a postgraduate research program, and in an environment in which a one-to-one working relationship has had to last for several years.

A concise description of the role of the supervisor and the role of the student would be a particularly useful point of reference for both parties, so that each could try to understand the other's perspective. However, the difficulty with this is that there are as many perceptions of the specific role of a postgraduate research supervisor as there are postgraduate research supervisors and, equally, the specific roles of students vary depending upon their capabilities and potentials, as well as the specifics of the research environment.

The traditional relationship between the supervisor and student was often based upon a master and apprentice model. In such an arrangement, the apprentice researcher followed behind the master, conducting specified research tasks, and observing the master's research characteristics so that, ultimately, the apprentice might also become a master in the same mold as the original.

In recent decades, there has been a significant transition from the traditional research arrangements, which were conducive to a master/apprentice model, to a broad range of new postgraduate models. These arrangements can include:

- Industry-based or industry-oriented research.
- Credit-point-based professional research degrees, which include both coursework and research.

- University-specific arrangements, such as higher Doctorates, or Doctorates by publication.

Additionally, some postgraduate programs are structured to enable senior industry professionals to upgrade their knowledge and skills in particular areas. The result is that, in modern research environments, the *apprentice* could well be a senior industry professional who:

- Is older than the actual supervisor.
- Has more overall professional experience (if not field-specific research experience) than the supervisor.

Hence, maintaining the traditional master/apprentice model becomes difficult, to say the least.

Research supervision has not only changed in nature because of the changing research environment but also because of the increasing number of people undertaking postgraduate research programs. The first of the modern research-based PhDs was awarded in the 19th Century, and for the remainder of that century, the number of people annually involved globally in postgraduate research could have been measured in the hundreds. In the early decades of the 21st Century, the annual numbers are measured in the hundreds of thousands. A corollary of increasing participation rates, and the move by universities around the world to achieve greater efficiencies, is that the nature of research supervision has also changed. Whereas a 19th Century *master* may have only had one postgraduate *apprentice*, the 21st Century *master* may have 10–20 *apprentices*.

As each research *master* increases the number of research *apprentices*, in line with increasing university and societal expectations, the breadth of the *master's* knowledge tends to increase but the depth, in specific areas, decreases. A modern *master* needs to be an expert in:

- Research grant applications.
- Occupational health and safety.
- Finances and budgeting.
- Quality processes,

as well as a whole plethora of other peripheral activities. The time that is genuinely available to allocate to each *apprentice* is therefore diminished. As a consequence, modern *apprentices* need to be more independent and proactive than those of the past and it is also not uncommon for modern *apprentices* to achieve a much greater depth of knowledge in their specific field of study than their *master*.

Given the changes that have occurred in the relationship between the supervisor and the student, as a result of the changing nature of research, one may well be tempted to ask if a simple definition for the role of the supervisor can be still be put forward. However, it is important to note that, although the techniques and practices of the supervisor may have changed over the course of centuries, there is still one fundamental attribute of research supervisors that remains unchanged - that is,

> *To guide and mentor students in such a way that they can learn about the systematic processes of discovery...*

In this chapter, we examine the different approaches to guiding and mentoring that have become commonplace; examine how these can lead to conflict and, subsequently, what can be done to minimize conflicts.

4.2 Supervisor Types

In order to make a discussion on the relationship between the supervisor and student tractable, in this text, supervisors are simplistically divided into two basic categories:

- Master/Apprentice.
- Laissez-Faire.

Needless to say, there are numerous shades of grey in between the two extremes put forward here, but the simplistic grouping serves a useful purpose for discussions.

Each of the supervisor categories, at the extremes, tends to maintain a particular belief and value set that is used to govern the approach which they adopt in postgraduate supervision. A range of some commonly-held beliefs is suggested in Table 4.1. Novice supervisors may be able to identify the grouping into which their beliefs best fit.

Not all research students will be able to fit in with one or other of the two basic supervisory categories. It is important for the supervisor to understand where he/she fits and what sort of attributes a potential research student will find best suited to his/her personality. If there is no match between supervisor and student expectations at the outset of the program, then clearly there are going to be difficulties along the way. At the very least, both parties need to be given an opportunity to air their views and expectations to see if a mutually beneficial partnership can be formed.

Table 4.1 Common supervisor beliefs

Issue	Master/Apprentice Supervisor's Beliefs	Laissez-Faire Supervisor's Beliefs
The Research Student	Needs to be carefully trained by an expert	Is already qualified and capable of self-learning
The Research Program	Needs to be mapped out by an expert in order to avoid mistakes	Needs to be mapped out by the student as part of the learning process
Independence	Something that is earned after the apprenticeship	An integral part of the learning process
Publication	The master's name should always appear first because the apprentice is only an assistant in the program	The student's name should always appear first because the student is the driving force and the supervisor is the guide
Interaction	Daily interaction and discussion	Weekly or monthly interaction
Supervisor's Knowledge	Should be far greater and deeper than the student's in the specific field of research	The student may have more depth in the specific field but the supervisor has broader overall knowledge about the process of research

Traditionally, students and supervisors sought each other out, based on mutual research interests. However, it is increasingly the case that students and supervisors are brought together by *the system*. For example, in a collaborative research project, a research student may be recruited in the same way as any other employee would be. Sometimes this could occur through a selection panel rather than through the explicit choice of the supervisor – and sometimes against protestations from the potential supervisor.

It is therefore important that the supervisor and student both have a good grasp of the advantages and disadvantages of each particular type of supervisory arrangement, in order to cope with inevitable problems that will arise.

Some of those who study Table 4.1 will say that it clearly portrays the *Laissez-Faire* supervisor in the best light, while others will say that the reverse is the case. Some research students view postgraduate research as an opportunity to study with an eminent expert and feel that, by walking in his/her footsteps, they can acquire some of the attributes of that expert. These students may actually prefer a *Master/Apprentice* relationship. Other research students view postgraduate research as a means of fulfillment through self-learning and innovation. These students may prefer a *Laissez-Faire* relationship.

From the supervisor's perspective, there is also the issue of student capability. A supervisor who genuinely believes in the *Laissez-Faire* approach may need to recognize that a research student who is put in their care simply doesn't have the capacity or independence to act effectively in an independent learning environment. This is especially true when research students are recruited from countries where undergraduate learning programs are almost entirely rote-based, with little latitude for independent thought and development.

It is important to note that neither of the supervisor belief sets are perfect in themselves. Each has advantages and disadvantages. A suggested set of comparative attributes is put forward for consideration in Table 4.2, in relation to factors that influence the outcomes of the research program and the performance of the student.

In summary, the *Master/Apprentice* approach to supervision may provide a more systematic approach to imparting the rigors of research to the student. The key disadvantages are that the student becomes dependent upon the supervisor and inherits both positive and negative research traits – overall, the potential for high creativity is reduced by the process. In the *Laissez-Faire* approach, the obvious advantage is that what the student learns through experiential self-learning may be far more powerful as a research tool than rote learning derived from drilled research procedures. The disadvantages are that the overall risk of presenting research, which is unacceptable to peers, may be higher, and that the student may develop lax or incorrect research practices without sufficient supervisory oversight.

Table 4.2 Possible advantages and disadvantages of supervisor types

Factor	Master/Apprentice Relationship	Laissez-Faire Relationship
	A = Advantage *D* = Disadvantage	
Risk	*A* More likely to lead to research outcomes accepted by peers	*A* Higher probability of achieving an outstanding outcome in one's own right
	A Research project plan/theory developed with high level of expertise	*D* Possibility of an outcome which is not accepted by peers
	D Lower probability of achieving an outstanding outcome in one's own right	*D* Research project plan/theory may have fundamental flaws or gaps
Learning	*A* Experience comes from disciplined training rather than unguided self-learning	*A* Experiential self-learning may lead to student becoming far superior to supervisor
	D Student only strives to reach the level of the supervisor	*D* Mistakes may not be identified during the program

(*Continued*)

Table 4.2 Continued

Factor	Master/Apprentice Relationship	Laissez-Faire Relationship
	A = Advantage	
	D = Disadvantage	
Originality/ Creativity	*A* High levels of drilled rigor may lead to good research practice even without creativity	*A* High level of creativity is promoted
	D Creativity stifled – student tends to become a clone of the supervisor	*D* Creativity divorced from rigor may not lead to good research practice
Theory	*A* Generally checked with a high level of expertise	*A* Student more likely to self-check work and get alternative opinions
	D Student may be too trusting of supervisor's erroneous advice	*D* Fundamental gaps or flaws may slip through undetected
Research Method	*A* Student develops a good research technique if supervisor is good	*A* Student's self-learning may lead to deeper experiential learning and high-level research skills
	D Student may develop a bad research technique if supervisor is bad	*D* Student may not develop rigorous techniques without drilling

4.3 Interaction between Students and Supervisors

The purpose of the interaction between the research student and the supervisor is to:

(i) Plan a research program and modify the plan according to changing requirements.
(ii) Ensure that research takes place in a safe, ethical, systematic and rigorous manner.
(iii) Assist in the development of a series of ideas into a cohesive structure that can become a major or minor thesis.
(iv) Provide ongoing, constructive feedback to the research student in relation to the research program.
(v) Correct or resolve research problems arising from the research plan or conduct of research.
(vi) Assist in the publication of ideas; the peer review of ideas, and the analysis of peer reviews.
(vii) Facilitate the purchase of equipment and resources required for the conduct of the research.
(viii) Facilitate interaction with governing bodies of the university and/or outside collaborators.

4.3 Interaction between Students and Supervisors

The relationship between a research student and supervisor also has other professional facets, and the specific interaction between supervisors and students varies substantially, according to the working relationship between the two, and according to the alignment of personal beliefs.

It is self-evident that the productivity of a research student and supervisor is greater when the relationship between the two is good. Apart from the obvious human tendency to consciously or subconsciously avoid people with which one does not have some personal affinity, there are other professional issues that also come into play. In particular, when there is a good working relationship, there is also a more honest exchange of ideas and opinions than there is in a purely formal professional relationship.

A key problem here is that when students and supervisors have no personal affinity, then the relationship moves on to a formal professional footing. This tends to lead on to rigid meetings, with typical outcomes being that:

- The supervisor asks the student to achieve particular results, with which the student may not agree.
- Rather than causing a dispute in a formal meeting, the student agrees to particular courses of action and then does the minimum amount of work necessary in order to appear to have cooperated.
- The supervisor becomes aware that the student may be acting in a passive-aggressive manner but, rather than exacerbate an uncomfortable working relationship, tends to accept the outcomes.

A hallmark of rigid relationships between the supervisor and the student is that the student tends to appear at meetings with formal statements of milestones that have been achieved and of the *correct answers* to the research tasks that have been set. Supervisors then take on a position of infallibility and students take a position built on defending or excusing their fallibility.

Such meetings inevitably lack in-depth discussion on what has not been achieved or of the *incorrect answers* that resulted from the research – that is, the reality of the situation. The end result is that the interaction between the student and the supervisor is little more than a series of formalities at which a superficial exchange of project management information is enacted. The same results (or better) might equally be achieved simply by an exchange of electronic correspondence, and the intrinsic value of a personal meeting has been lost.

It is important that a supervisor not present himself/herself as an infallible font of knowledge that cannot be challenged by the student. It is beneficial for both the student and the supervisor to freely challenge each other's views without rancor or bitterness. As Lao Tzu wrote in the *Tao te Ching*, two millennia ago,

"The Master doesn't glitter like a jewel but is as rugged and common as a stone."

In other words, a true leader (master) presents all his facets to his followers, not just the polished ones.

In situations where there is a good working relationship between the supervisor and the student, there tends to be more in-depth discussion on what has gone wrong with the research:

- Why have experiments failed?
- Why have expected milestones have not been achieved?
- What likelihood there is of achieving particular results in the future?

These discussions are an invaluable part of the research learning process – which needs to examine and understand the *wrong answers* as much as the *right answers*.

In a good working relationship, formalities are removed; students feel comfortable talking about problems – supervisors feel comfortable with the idea of talking about their knowledge limitations. Both accept the concept of personal fallibility and are prepared to acknowledge possible mistakes in a positive fashion.

Putting the situation in another way, when there is a poor working relationship between a supervisor and a student, then research integrity tends to suffer in the process. Student morale drops and supervisor interest in the research diminishes, thereby creating a downward spiral. Even if a good professional relationship can still be maintained under such circumstances, the likelihood of positive outcomes is diminished because the frankness of discussions is eliminated through formality. Needless to say, if both the working relationship and the professional relationship are poor, then the likelihood of a research student achieving a positive outcome is severely limited.

The major problem in supervisor/student relationships is that, often, neither the student nor the supervisor can have absolute freedom in their choice of working colleagues. Students and supervisors are frequently brought together through the machinations of university departments and of common research fields, rather than through the intangible factors that may create a genuinely good working relationship.

It would be all too easy to suggest that when both the working relationship and the professional relationship between a supervisor and a student have eroded, then both parties should seek to make alternative arrangements, through a higher level authority within the university system. However, limitations of expertise in particular fields often restrict the supervisory choices available to students and, in any event, supervisors have a professional responsibility to always be the *adult in the room* and make any given set of circumstances work effectively, whether they like it or not. That is the burden of professionalism – when personal working relationships fail, a good professional relationship still has to be maintained.

4.4 Typical Supervisor/Student Problems

In this section, the objective is to examine the sorts of problems that arise between supervisors and research students during the course of a postgraduate research program. Perhaps, by understanding the nature of these problems, and recognizing that they are relatively commonplace in many research projects, supervisors may come to appreciate that most can be resolved through goodwill or, sometimes, through a self-disciplined approach to professional practice. In many instances, simply recognizing that supervisor/student problems are a natural consequence of a long-term, one-to-one relationship between individuals, cast together through technical circumstance, assists in developing a more mature approach to tackling the solution.

In Table 4.3, a range of typical supervisor/student problems are categorized into broad groups, to highlight the volatile and complex nature of the relationship that can often exist in a postgraduate research program. The problems are cited in terms of the research student's perspective so that supervisors can get an insight into the challenges they face.

The first point to make about the sorts of problems that are cited by students is that they can stem primarily from supervisor arrogance – and a lack of professional maturity in managing people. It is often (mostly) the case that research student supervision is the very first people-management task that a supervisor has undertaken as a professional, so it is not surprising that there will be mistakes and problems along the way. Sometimes, unfortunately, supervisors don't even realize that their students feel ill-disposed towards them and harbor hostility and resentment.

Problem Groups (i)–(ii) can often arise when a supervisor underestimates the intelligence of the research student and their capacity to see through self-serving arguments made by the supervisor. The supervisor needs to be on the high moral ground before there is even a chance of tackling these issues. If a supervisor has:

- Underestimated the student.
- Failed to create a professional working relationship.
- Failed to put sufficient work into the actual supervision,

then, clearly, resolving these issues is going to be difficult.

88 The Relationship between the Supervisor and the Student

Table 4.3 Typical supervisor/student disputes (student perspective)

	Problem Group	Typical Problems (Student Perspective)
(i)	Interpersonal	• Supervisor doesn't like me • I don't like my supervisor • Supervisor and I argue over everything • Supervisor thinks that I lack intelligence
(ii)	Belief-system/ Alignment	• Supervisor is arrogant • Supervisor won't allow me to do the project in my own way • Supervisor isn't doing his/her job • Supervisor doesn't contribute anything • Supervisor doesn't know anything about the specifics of the subject
(iii)	Technical	• Supervisor and I completely disagree with the plan for research • Supervisor will not accept my findings – asks me to repeat experiments
(iv)	Ethical/Moral	• Publications – whose name should be first • Supervisor has presented my work as his own research • I have published a paper without my supervisor's name • Supervisor has made me publish the same paper in two different journals with different titles • Supervisor asks me to falsify results • Supervisor is deliberately delaying my research so he can use me as a publishing machine for his benefit • Supervisor insists on borrowing other research work without providing due credit

Unpleasant though it may be, at some point, a supervisor will need to engage in some introspection and decide whether he/she is genuinely at fault. And, it also needs to be kept in mind that, being the *adult in the room*, the supervisor is always technically at fault in the relationship even when a student is:

- Cantankerous.
- Uncooperative.
- Slack in his/her work practices.

The role of people management and supervision is to prevent these sorts of behaviors from getting out of hand in the first instance.

Problem Group (iii) should potentially be the most straightforward one to manage through. If there are technical disagreements between the supervisor and the student, then they should be resolved in a rigorous technical way, and without resort to personal issues or confrontation. Sometimes, the simplest way of resolving technical disputes is to seek the opinions of other peers in the field – as a circuit breaker. Once each party has articulated his/her arguments, there is little to be gained by repeating them, over and over, with increasing animosity. If an agreement cannot be reached between the parties additional, external input is clearly required.

Problem Group (iv) represents the sorts of issues that create the greatest conflict, and usually escalate well beyond the realm of the supervisor and the student. These issues arise because either the student or supervisor views the other party as unethical or immoral. It is often difficult to know how to proceed in such matters, save to consider:

- Is the supervisor certain that he/she has acted in a completely ethical manner?
- Are the supervisor's actions ones that would be considered ethical, as perceived by an outsider who is an independent peer in the field?

If the supervisor's answers to one or other of these questions is no, then there are real problems – often ones that would require referral of the entire matter to a higher level authority within the university. These issues are covered in detail in Chapters 9 and 13.

The obvious answer to tackling the Group (iv) issues is to say that they should not arise in the first instance because, if they have, then a supervisor has made grave errors and has failed in his/her duties. However, there is also the possibility that a student has misinterpreted various actions – perhaps there has been a breakdown of communications. The supervisor has to act to resolve this quickly and professionally.

At all times, it is imperative that a supervisor does not retaliate to student animosity with further animosity. The supervisor's role is to dampen problems that occur, not to inflame them in any way. And, needless to say, there is no room on the part of the supervisor for any form of hostility (verbal or otherwise) or retaliation.

In Section 4.5, a number of techniques for avoiding a terminal breakdown in the supervisor/student relationship are put forward.

4.5 Conflict Management and Resolution of Disputes

One of the most important aspects of any professional relationship between individuals stems from the old adage which simply states that,

> *"Before putting someone in their place, try to put yourself in their place."*

In other words, everyone has a different perspective and, before taking a counter perspective, it is particularly important to understand that of the opposing party.

Another important aspect of conflict resolution is to avoid conflicts in the first instance. Disputes in academia should not become a war of attrition. This is particularly true in postgraduate research supervision where one needs to understand that a relationship must generally continue after the dispute has ended. Hence, both the supervisor and the student need to consider what will happen after a dispute, and what the consequences may be for the research program itself.

A good underlying rule for managing professional disputes is therefore to draw a flowchart – either mentally or physically – on how the process will pan out from the first encounter. Specifically, consider:

- What happens if one party or another inflames the situation?
- What happens if other colleagues or more senior management are brought in by the research student to resolve the dispute?
- Whose side will management take when the relationship between supervisor and student has broken down – and it is the supervisor's responsibility to ensure that it doesn't?
- Regardless of the intermediary steps in the chart, the end result may be that the supervisor and student will still need to work together – in a positive and productive way – if good research outcomes are to be achieved.

As a starting point, in examining conflicts that arise between supervisor and student, it needs to be understood that the role of people management is not simply about telling others what to do. Professional management involves the following basic activities:

- Understanding the capabilities and predispositions of the other party.
- Determining the willingness of the other party to undertake a task to the best of their ability.
- Negotiating tasks and responsibilities with the other party in such a way that the other party becomes a willing and enthusiastic participant in the task rather than a begrudging servant.
- Enabling the other party to undertake a task (i.e., providing the resources necessary for success).
- Ensuring that the other party is fully briefed on the boundaries of the task execution (i.e., what is permissible and what is specifically disallowed).

4.5 Conflict Management and Resolution of Disputes

- Monitoring/overseeing the efforts of the other party on a progressive basis (i.e., not just at the end when it is evident that the task has/hasn't been completed).
- Providing additional support/input when it is clear that the other party's execution of the task is falling behind.
- Understanding whether a failure to complete a task is the result of a lack of capability, resourcing, support, willingness or work ethic.

As a supervisor, who may be entering into a conflict/dispute with a research student, it is important to undertake a self audit, and ensure that all of these basic requirements for management have been fulfilled before even considering the actual dispute itself. From the supervisor's perspective, it may also be helpful to indulge in some introspection about how well he/she can deal with conflict and resolution.

In 1974, Thomas and Kilmann *(Thomas and Kilmann, 1974)* published a book outlining a conflict resolution instrument (the *Thomas-Kilmann Conflict Mode Instrument – or TKI*) to help people better understand where they were placed in terms of their ability to deal with conflict.

The TKI identifies a person's behavior in a conflict situation, by looking at two dimensions *(Kilmanndiagnostics.com, 2015)*:

- *"Assertiveness – the extent to which the person attempts to satisfy their own concerns.*
- *Cooperativeness – the extent to which the person attempts to satisfy the other person's concerns."*

Figure 4.1 shows the various, possible TKI conflict handling modes, mapped across these two dimensions.

Thomas and Kilmann described the various approaches to handling conflict in the following way *(Kilmanndiagnostics.com, 2015)*:

> *"These two basic dimensions of behavior define five different modes for responding to conflict situations:*
>
> 1. **Competing** *is assertive and uncooperative – an individual pursues his own concerns at the other person's expense. This is a power-oriented mode in which you use whatever power seems appropriate to win your own position – your ability to argue, your rank, or economic sanctions. Competing means "standing up for your rights," defending a position which you believe is correct, or simply trying to win.*

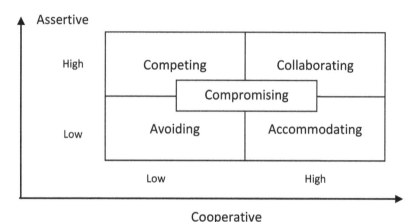

Figure 4.1 TKI conflict handling modes.

2. **Accommodating** is unassertive and cooperative – the complete opposite of competing. When accommodating, the individual neglects his own concerns to satisfy the concerns of the other person; there is an element of self-sacrifice in this mode. Accommodating might take the form of selfless generosity or charity, obeying another person's order when you would prefer not to, or yielding to another's point of view.
3. **Avoiding** is unassertive and uncooperative – the person neither pursues his own concerns nor those of the other individual. Thus he does not deal with the conflict. Avoiding might take the form of diplomatically sidestepping an issue, postponing an issue until a better time, or simply withdrawing from a threatening situation.
4. **Collaborating** is both assertive and cooperative – the complete opposite of avoiding. Collaborating involves an attempt to work with others to find some solution that fully satisfies their concerns. It means digging into an issue to pinpoint the underlying needs and wants of the two individuals. Collaborating between two persons might take the form of exploring a disagreement to learn from each other's insights or trying to find a creative solution to an interpersonal problem.
5. **Compromising** is moderate in both assertiveness and cooperativeness. The objective is to find some expedient, mutually acceptable solution that partially satisfies both parties. It

4.5 Conflict Management and Resolution of Disputes

falls intermediate between competing and accommodating. Compromising gives up more than competing but less than accommodating. Likewise, it addresses an issue more directly than avoiding, but does not explore it in as much depth as collaborating. In some situations, compromising might mean splitting the difference between the two positions, exchanging concessions, or seeking a quick middle-ground solution.

Each of us is capable of using all five conflict-handling modes. None of us can be characterized as having a single style of dealing with conflict. But certain people use some modes better than others and, therefore, tend to rely on those modes more heavily than others – whether because of temperament or practice."

It is useful to understand one's own underlying predispositions to various conflict resolution approaches before attempting to resolve a serious dispute with a research student.

Disputes in postgraduate research can often arise from opposing parties taking intransigent views on a particular subject, without sufficient understanding of why the other party has a counter view. Moreover, once intransigent positions have been locked in and have been communicated, both parties in a dispute can unfortunately push the situation into a win-lose context, in which one party has to withdraw in embarrassment. The issue in supervisor-student conflicts is that the supervisor is in a position of authority and may win a conflict battle in the short term – but at what cost?

The only long term victory that a supervisor can ever truly win is the ongoing respect and cooperation of the research student to complete the research program to the highest possible standard. Everything else constitutes a shallow victory.

Chapter 9 of this book will deal with the issue of conflict resolution in more detail.

5

Understanding the Research Environment

5.1 Overview

Research supervisors, as a result of their own undergraduate and postgraduate experiences, should already have a good working knowledge of the university and the encompassing university system. However, the overall picture is very complex and, quite often, those starting out in academic/research positions have not been fully exposed to the complete structure of the university or the system in which it operates. For these reasons, it is worthwhile to cover/review the basic elements that exist to support academic/research activities.

One of the problems in covering these issues is that each nation/region has its own distinct set of operating principles for universities and, within those national/regional principles, each university tends to have its own structures and procedures. Nevertheless, there are sufficient commonalities – or, at least comparable structures – within the global university system to make this examination worthwhile.

Understanding the university/research environment is important in the context of supervision because, beyond its academic/research constructs, a university is a complex organization with complex business structures and processes that are essential in order to enable it to function. Notwithstanding the higher educational purpose to which supervisors need to aspire, it is important to recognize that these constructs and responsibilities are a consequence of being in a large organization which has responsibilities for:

- Funding/finances which are, at a minimum, in the hundreds of millions or, at the larger end, billions of dollars.
- The safety and wellbeing of thousands of staff and tens of thousands of students.
- Reporting and accountability to funding agencies, governments, etc.

Moreover, universities are a global phenomenon and, while each operates in its own nation and region, the reality is that its performance needs to be

96   Understanding the Research Environment

considered relative to global best practices, rather than just parochial measures and standards.

5.2 A Perspective on Global University Numbers

A common question that many academics and students have relates to how many universities there are in the world. The answer to this question depends upon the specific definition of the word *university* and, with numerous possible interpretations, unsurprisingly, the number of reported institutions also varies.

One of the most extensive evaluations of tertiary institutions is undertaken by the Cybermetrics Lab, a research group belonging to the *Consejo Superior de Investigaciones Científicas* (CSIC) in Spain. This group creates the Webometrics Ranking of World Universities and includes some 24,000 institutions in its assessment of university web activities *(Webometrics.info, 2015)*. These 24,000 institutions include research intensive universities; teaching-only universities; institutes of technology and community/technical colleges.

The Shanghai Jiao Tong University's Academic Ranking of World Universities (ARWU) is a respected assessment tool which examines the relative performance of universities around the world. However, the ARWU only examines universities which have a significant element of research activity, according to the following criteria *(Shanghairanking.com, 2015b)*:

> "ARWU considers every university that has any Nobel Laureates, Fields Medalists, Highly Cited Researchers, or papers published in Nature or Science. In addition, universities with significant amount of papers indexed by Science Citation Index-Expanded (SCIE) and Social Science Citation Index (SSCI) are also included. In total, more than 1200 universities are actually ranked and the best 500 are published on the web."

So, for the purposes of those undertaking postgraduate research supervision, the answer is that there are nominally around 1,200 universities in the world which have a significant element of research, and tens of thousands of institutions which operate at a tertiary level but have no research (or limited research) activities.

The total number of postgraduate research students globally varies significantly from day to day, as new universities are formed, or older ones broaden from teaching activities into research. Suffice to say that there are hundreds of thousands of active postgraduate research students at any one time, enrolled in degrees ranging from Master's to Doctoral, to Higher Doctorates, Industry Doctorates, Doctorates by Publication, and so on.

The number of tenured academic positions available annually in research-active universities is also difficult to determine accurately, but is low relative to annual postgraduate graduations. The implications of this are profound for the research supervisor's approach to his/her task. Specifically, because there is a large annual shortfall in tenured academic positions relative to the number of postgraduates who may ultimately wish to fill them – and the shortfall is annually cumulative – the likelihood of a postgraduate student achieving a tenured university academic position in a research-active institution is likely to be very low. It is important for supervisors to keep this in mind in terms of what sort of skills will be relevant to the postgraduate student for his/her future career.

5.3 International Rankings

5.3.1 General Issues

Postgraduate research supervisors need to understand that their performance is not only being scrutinized by their direct line managers, faculties and universities but also, increasingly, as part of a global assessment of universities and their performance.

In recent years, a number of international assessment schemes have been put into place in order to determine the relative performance of universities at a global level. Needless to say, the performance of an institution can vary significantly depending upon the assessment parameters/metrics that are deployed in the ranking scheme.

The major rankings which are deployed in universities include the:

- Shanghai Jiao Tong Academic Ranking of World Universities (ARWU).
- Times Higher Education (THE) University Rankings.
- Quacquarelli Symonds (QS) University Rankings.
- Leiden University Centre for Science and Technology Studies (CWTS) Rankings.
- Consejo Superior de Investigaciones Científicas (CSIC) Webometrics Rankings.

Each of these has a different focus and, of course, some institutions perform better on some rankings than others. For example, the Webometrics Rankings look primarily at an institution's web presence and content. The Leiden

Rankings examine university publications on the Thomson Reuters *Web of Science* database *(Thomson Reuters, 2015)*.

In general, university rankings tend to focus on research activity rather than teaching activity. Teaching is inherently difficult to measure, and statistical data that is derived from information such as:

- Pass-rates
- Attrition-rates
- Student-satisfaction surveys,

can all be interpreted in a range of different ways. For example, high attrition-rates and low pass-rates can arise because an institution has rigorous and demanding educational programs, or because the institution has poor educational attributes. A high student satisfaction result can arise because an institution has lax teaching requirements or because it has outstanding educational staff.

For these reasons, international rankings agencies tend to focus upon research outcomes, where there are numerous, semi-objective research metrics available in order to provide relative measures of institutional performance. Despite these measures, there is always debate about which measures should be included or excluded, and what weighting each measure should have when it is included. So, there is seldom universal agreement on rankings methodologies, and one needs to treat each of them with caution because they each have strengths and limitations.

University rankings schemes lose their significance, however, if:

- The assessment methodology changes from year to year or is generally inconsistent.
- The metrics deployed in the rankings include subjective assessments of institutions (e.g., perception surveys).
- Institutions can readily *game* the rankings to achieve rapid improvements in their relative positions.

With these points in mind, one set of assessments which provides breadth, consistency, objectivity and intrinsic inertia (which reduces gaming of the system), is the Shanghai Jiao Tong Academic Ranking of World Universities (ARWU). For these reasons, the ARWU is used extensively as an indicator of overall institutional performance in a relative context.

The ARWU assessments of institutions are based upon four criteria and six indicators, as shown in Table 5.1 *(Shanghairanking.com, 2015b)*. The notable attributes of the ARWU are its:

5.3 International Rankings

Table 5.1 ARWU ranking criteria (Shanghairanking.com, 2015b)

Criteria	Indicator	Weight
Quality of Education	Alumni of an institution winning Nobel Prizes and Fields Medals	10%
Quality of Faculty	Staff of an institution winning Nobel Prizes and Fields Medals	20%
	Highly cited researchers in 21 broad subject categories	20%
Research Output	Papers published in Nature and Science*	20%
	Papers indexed in Science Citation Index-expanded and Social Science Citation Index	20%
Per Capita Performance	Per capita academic performance of an institution	10%
Total		100%

- Focus on high-end research – publications in prestigious journals and with high citations.
- Intrinsic inertia to prevent short-term gaming – an institution needs to perform well in terms of alumni achieving international distinctions – this needs to be achieved over decades.
- Focus upon globally transformative research – as indicated by the number of awarded Nobel Prizes, Fields Medals, etc.
- Focus on engineering, medicine, science and technology rather than humanities.

The ARWU favors long established institutions because, in order to perform at the high end of the rankings, an institution needs to have a demonstrated track record of having created a world-class alumni.

The alumni metric within the ARWU is therefore important in the context of the inertia it provides to prevent *short-term gaming* of the system, but also because alumni are an indicator of teaching performance. If an institution is to be deemed to be the best in the world, then it needs to demonstrate that its graduates have achieved the international pinnacle of their respective fields.

The strengths of the ARWU approach are reflected in the top-end rankings of institutions. The limitations of the ARWU are reflected in the middle and lower ends of the ranking scale, because the metrics that are employed are far too coarse for small (or newly-established) institutions to interpret. For example, metrics pertaining to the number of Nobel Prizes or Fields Medals are not particularly helpful to the majority of institutions around the world which would have no entrants in these fields.

There is also the issue that ARWU doesn't necessarily consider newer prestigious awards – for example, the Kavli Prize for scientific research *(Kavliprize.org, 2016)* and The Breakthrough Prize for fundamental physics, life sciences and mathematics *(Breakthroughprize.org, 2016)*. It also needs to be noted that the Nobel Prize is not awarded across a broad range of research categories – for example, biology is excluded.

The problem with updating ranking scheme metrics from year to year to reflect emerging trends in academic/research achievement is that such changes create inconsistencies when comparing institutional rankings across the years. The choice then becomes one of consistency or relevance – and generally consistency wins out, albeit meaning that those using the rankings need to understand the intrinsic limitations.

Nevertheless, the implications of the ARWU and its broad international acceptance are profound for research supervisors. The fact that all research-intensive institutions are assessed by ARWU metrics means that research supervisors need to focus on quality and not quantity. That is, quality of published work, quality of graduates, etc. There is little to be gained from supervisors simply trying to increase their metrics, because the over-arching assessment of institutions is on quality with a very high international benchmark.

5.3.2 How International Rankings Affect Research Supervisors

There are numerous points which should become apparent from any investigation of university rankings, specifically:

- The highest ranked institutions are those which have had sustained contributions to transformational change at a global level, over decades or centuries – this is generally measured in internationally renowned alumni; major breakthrough discoveries, etc.
- The world's most highly ranked institutions are those with enormous financial and physical resources, and generally large academic/research staffing levels.
- The highest caliber academic staff are naturally drawn to the highest ranked institutions.
- Highly ranked institutions draw the best staff and students from all over the world, rather than from just their own local regions.

- High caliber staff and students tend to push each other upwards to achieve sustained levels of excellence.
- The bulk of the global resources required in order to become a world leading institution (e.g., Nobel laureates, Fields Medalists, etc.) are already located within the world's leading institutions – so university rankings become self-fulfilling, and the order is difficult to change because of intrinsic inertia in the system.
- It is generally not feasible for small, poorly-resourced institutions to make dramatic improvements in their rankings because of the systemic inertia – and the fact that improvements to rankings generally require large injections of funds and resources across a wide array of subject areas.

A research supervisor therefore needs to be aware of the constraints that are imposed by the institution in which he/she is operating. For a fortunate few, there will be the luxury of working in a world leading institution, and the ability to attract the highest caliber research students from around the world. For all the other supervisors, an assessment needs to be made of what it is feasible to achieve in the existing environment considering:

- The number and caliber of the research students which the institution is able to attract.
- The resources and technical support within the institution.
- The intellectual caliber/environment of the institution in which the research student will work (will it inspire the student upwards or drag the student downwards?).

Clearly, an individual research supervisor will have little or no control over the institutional setting. However, this should not be used as a rationalization for tolerating under-performance. The university environment is global, and a research supervisor always needs to think of himself/herself as part of a global research effort, and not just part of one isolated institution. Supervisors therefore need to collaborate with world leaders in the field if their research is to be meaningful in an international context. Such collaborations don't need to be costly and neither do they necessitate travel. At the very minimum, they can be undertaken through internet-based technologies.

To this end, each and every supervisor needs to be looking at his/her supervision and research relative to the world's best players in the specific field – and not relative to internal standards within his/her home institution (unless of course that institution is an independently recognized world-leader in that field).

Part of a supervisor's role is to monitor the activities (i.e., publications, theses, etc.) of the world's best institutions, in his/her field, and ensure that, notwithstanding obvious resource constraints, outputs from the home institution measure up favorably.

There are many aspects of postgraduate research which are not constrained by resources, including:

- Academic rigor.
- Systematic approaches to investigation.
- Detailed, comprehensive analysis of data/information.
- Balance and impartiality in presentation of results.
- An accurate depiction of reality – including self-assessed limitations and strengths of presented research.

It also needs to be noted that resources are not the solution to every research problem, and that they can equally become an unforeseen impediment because they sometimes override the need for systematic and careful consideration of issues prior to synthesis or experimentation.

There may be any number of reasons why a supervisor's home institution cannot duplicate what exists in a world-leading institution. However, to do less than strive for world's best practice in a given field is to do a disservice to the research student.

5.4 Institutional Funding Arrangements

In any university, in any country in the world, there are only a limited number of possible sources for institutional funding, and so universities around the world tend to have similar funding models. The major differences between institutions are the relative contributions from each source. The sources include:

(i) Direct national/regional government institutional funding grants.
(ii) Student tuition fees.
(iii) Benefactorial donations.
(iv) Investment income from endowments.
(v) Commercial income from contract research and development; royalties; intellectual property (IP), asset sales, etc.
(vi) Competitive (merit-based) research grants from government agencies and private benefactorial organizations.

The key point to note from this is that, in general, although institutional funding may all be held in a central repository, the funding regimes for research and education can be treated separately in an accounting sense.

In most countries, universities aggregate funding sources (i)–(v) into a recurrent budget which is used to fund human and physical resources as per Figure 5.1. Competitive research grants, on the other hand, are funds that are *earned* by academic/research staff in a competitive national or regional process – and those funds are ear-marked for the specific projects for which funds were derived.

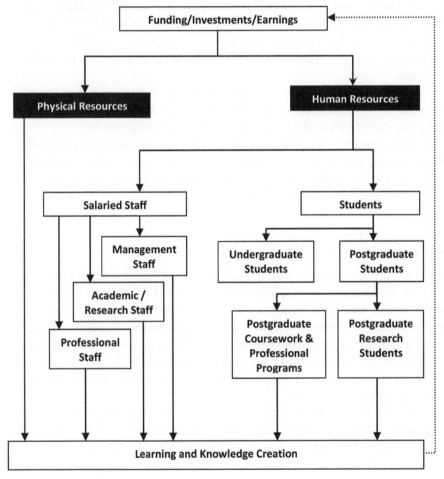

Figure 5.1 Basic elements of the modern university.

When academics/researchers are working on research projects funded by competitive research grants, they are effectively working in a specific cost-center for the university. However, research grants generally only cover the costs of new resources, new research staff (e.g., postdoctoral researchers), and postgraduate research students. They often do not cover the costs of the regular academic staff who are also working on the research. For these reasons, a university often has to cross-subsidize (i.e., *leverage*) funds from competitive grants with its own recurrent funding in order to make many research projects viable in cost-centered accounting terms.

When a university *leverages* its research funding, it is effectively transferring recurrent funds, which would normally be used for undergraduate and graduate education, to specific research projects. This is often referred to as an *in-kind* contribution to a research project, because academic/research salaries and technical support/infrastructure costs are awarded to a specific research cost center.

For this reason, it is particularly important that all academic staff, including those purely involved in research, understand that they are not isolated research entities but, rather, part of the larger university educational process.

In many countries, the funding provided for specific research projects, through competitive grants, is devolved through two primary entities/agencies – one for medical/health research and one for all other areas of research. For example, in the United States, the two agencies that provide these funding mechanisms are the:

- National Institutes of Health (NIH) *(NIH.gov, 2015)*.
- National Science Foundation (NSF) *(NSF.gov, 2015)*.

One reason for this globally widespread anomaly is that it is often difficult to separate research in medical science from national health and wellbeing issues – hence, these two areas tend to be combined into a single funding model, and the remainder of research activities are funded through another model. Needless to say, in the modern world, this becomes somewhat muddled because there are researchers in scientific/engineering fields who impinge upon medical research, and medical researchers who are predominantly conducting scientific research.

In addition to government funding for research, each nation generally has a collection of private benefactors and benevolent commercial organizations who also fund university research projects – although these tend to be primarily in areas of medical science.

Government and private/commercial research grants can also be competitive in nature and so, by definition, there are winners and losers. In any given year, academics and researchers seeking to move their research forward can only do so when they are successful in their grant applications. The problem with this is that it makes it very difficult for institutions to maintain systematic programs of research investigation because they are subject to the idiosyncrasies of competitive research funding providers.

For these reasons, the world's leading research universities tend to be those that are able to maintain a high and consistent level of base-load research funding for staff, which is independent of external competitive grant processes. This sort of funding is typically provided from investment income derived from university endowments.

Invariably, those institutions which have the highest research endowments are those that are able to maintain the highest levels of internal research funding – and thereby become a magnet for leading research staff from around the world – and thereby become leaders in research.

There is a *chicken and egg* scenario in this process because, in order to attract large endowments from benefactors, an institution has to have a track record of research excellence over decades or centuries. And, of course, in order to achieve research excellence, a university needs to have an internal income stream from endowments. Hence, the global university system has considerable inertia, and the disparities in endowments from the world's leading universities to the world's poorest universities are immense.

Those institutions at the bottom of the international ladder therefore need to work extremely hard in areas of research which have a very low cost base in order to improve their lot in life. A task made all the more difficult because there are so many institutions around the world seeking to undertake low-cost research which has significant impact, and which may potentially boost the standing of the institution. The task is made even more onerous when one considers that many of the world's universities – from wealthiest to poorest – all tend to identify low-cost-base research as an efficacious approach.

5.5 Institutional Structures

Each university has its own unique structural attributes, but there are common elements which exist in all institutions – in one form or another. Figure 5.2 shows some of the typical elements. In addition to the academic elements shown therein, there are of course a large number of corporate elements

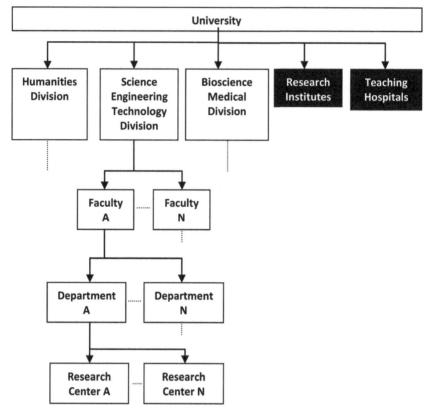

Figure 5.2 Typical elements in modern universities.

including personnel, finance and accounting, legal, information technology (IT), library, marketing, etc.

5.6 University Governance Structures

Universities, by definition, tend to adopt a collegial style of management in which significant decisions are made by groups of experts. This is partly for historical reasons and partly because universities have a very broad mix of products (i.e., courses, areas of expertise, activities, etc.) that are inherently difficult to manage by individuals working outside their area of expertise. Figure 5.3 shows the various elements typically found within an institution. The nomenclature for each position – particularly at the senior level – varies from country to country but the basic attributes are similar across institutions.

5.6 University Governance Structures 107

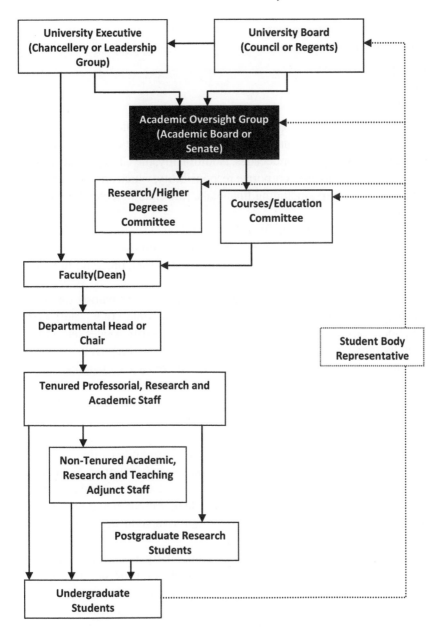

Figure 5.3 Typical university governance structure.

Universities are similar to many large organizations in the sense that they operate three levels of management:

- Strategic
- Tactical
- Operational.

The strategic group is composed of a Chief Executive Officer (CEO) and a board who act in concert to map out the long-term future directions of the institution over coming years or decades. Additionally, the CEO generally has executive oversight of areas of corporate responsibility, which are each headed by professional managers, including financials, information technology (IT), library, etc.

The tactical group in a university is that collection of senior academics that examines the long-term strategic directions of the institution (i.e., the *where*) and determines what specific actions need to be taken (i.e., the *how*) in order to achieve those objectives.

The operational group is that collection of academics whose focus is on managing structures that deliver short term outcomes (e.g., staff and resource allocations, degree completions, pass-rates, attrition rates, etc.) that incrementally move the university towards its long-term objectives.

In among these three management structures there are overlapping levels of functionality. Some strategic staff may do operational management and some operational managers (e.g., faculty deans and institute directors) may do strategic planning for their sphere of operations.

Of particular note in Figure 5.3 are the collegial elements of university governance. Specifically, these are a collection of committees that provide collegial management of a range of academic and research activities. Typically, these collegial bodies (e.g., academic board, higher degrees committee, etc.) are composed of representatives from each academic/research unit within the institution, and they serve as critically important oversight mechanisms for various aspects of education and research.

The collegial governance bodies within a university also provide consistency across the university in terms of standards, ethics, procedural fairness, etc. In some cases, they may also serve to provide a more efficient operating model for the institution – for example, preventing two departments from two different faculties each running a similar degree program.

From a research supervisor's perspective, the combined corporate and collegial structures that are in place in institutions have consequences. Firstly, a supervisor may ultimately be directly accountable to his/her *line managers* (e.g., departmental head, faculty dean, etc.) as well as to the collegial governance entities (e.g., higher degrees by research committee) for the supervisory process. Secondly, in the event of a dispute, a research student may have the option of by-passing a supervisor and his/her line managers altogether, and seek to have disputes resolved by the relevant collegial bodies.

5.7 Internal Performance Metrics

University governance bodies regularly scrutinize international rankings mechanisms to gain an insight into how universities are assessed in a relative sense. In addition, university management may also have performance obligations imposed by national/regional governments and other regulatory bodies. The sum total of these considerations will be devolved down to a series of performance metrics (either implicit or explicit) which will, by necessity, impact upon staff undertaking research supervision. In general, universities will have a statistician, or team of statisticians, constantly monitoring performance across all regionally, nationally and internationally recognized metrics – in order to understand how the institution is performing on a relative basis.

In a research context, there are numerous performance metrics that are in common usage. These include:

- Competitive research grant income.
- Current Doctoral student numbers.
- Doctoral completions.
- Research publication numbers.
- Research publication citation numbers.
- Publications in internationally esteemed publication vehicles (e.g., Nature).
- Patents and other intellectual property (IP) tools arising from research.
- Royalty income.
- Prestigious international awards/recognition (e.g., Nobel Prizes, Fields Medals, etc.).

High caliber research students have the potential to contribute to all these areas of performance. However, it is important to remember that a research student's task is to learn the process of systematic investigation and discovery – not to merely provide horsepower for an academic supervisor to improve his/her performance metrics.

One metric which is seldom considered is the gratitude that a research student has towards his/her supervisor after graduating from the university. This may manifest itself in large benefactorial donations to the institution if the alumnus becomes financially successful – and may prove to be of considerably greater value to the supervisor than any short-term gaming of internal performance indicators. It will certainly be of far greater value to the institution itself.

PART II

The Supervisory Process

6

Preliminary Tasks in Research Supervision

6.1 Overview

A novice research supervisor may be in a fortunate position where a research group leader decides to allocate one of his/her own students and projects to the novice to supervise. In these circumstances, some of the preliminary work of the supervisory process has already been completed. For most novice supervisors, however, they will need to deal with the basic tasks of:

- Definition/funding of research projects.
- Scholarship funding.
- Research student recruitment.
- Research candidature applications.

Although these are primarily bureaucratic/administrative processes, the reality is that they can end up consuming almost as much time as the academic aspects of the supervision. It is therefore important to examine the key factors that need to be considered in undertaking the basic initiating steps in the supervision, so they might be tackled in the most efficient manner.

6.2 Definition/Funding of Research Projects

Experienced, senior academics, with established research track records and reputations, may be in a position where they are constantly approached by final-year undergraduate students who wish to undertake a postgraduate research program with them. Additionally, a senior academic will have an established program of research and an existing research team that has well defined areas of research. For these academics, the early tasks associated with initiating a supervision are generally related to filtering of applicants, and deciding which ones will be the best fit for the research group.

For a novice supervisor, on the other hand, the onus is reversed and it is often the case that the novice has to make a value-proposition to attract a suitable postgraduate student. There are two possible options. The first, and simplest (but least strategic), is to allow a potential research student the luxury of defining his/her own research project. If the student is self-funded (i.e., has his/her own scholarship), and there are no significant physical resource requirements, then one can let the student build up some inertia in the chosen specialization, with the hope of using these research outcomes as the basis for future funding applications. The disadvantage of this approach is that a research student can potentially drive research activities in any self-preferred direction – and the nuances of that direction may not be compatible with what the supervisor sees as a long-term viable program of activity.

The second (more strategic) approach is to predefine the research project and thereby form the basis of a longer-term strategy for the supervisor to build a team. This, in turn, may involve funding of equipment, technical support, etc.

Sometimes, a novice supervisor will be allowed to build up a research activity within the research grouping of a more senior academic. This enables the novice to ride on the coat-tails of the senior academic and have a reasonable chance of securing competitive research funding by leveraging off the senior researcher's track record. However, for those who wish to go it alone, the task of building up research activity as a bootstrap operation is onerous. In order to be successful in securing competitive research funding from various national/regional or benefactorial funding agencies, one generally has to have a substantial track record. In order to establish a track record, one needs to have funds, so there is a circular problem to be tackled.

Universities often have competitive, internal funding mechanisms to support novice researchers in building up an area of research, through provision of minor equipment/support funding and potentially scholarship funds for postgraduate researchers.

Regardless of the starting point, it becomes evident that a novice cannot simply start the process of research supervision on a whim. Time and planning are required in order to systematically develop an area of investigation which can be resourced. It is also grossly unfair (if not unethical) to potential research students to have them commence their research program in the knowledge that there are insufficient resources available to complete that work.

In understanding the funding requirements for the project, a novice supervisor needs to be familiar with the overall corporate governance of the university. For example, it would be naive for a supervisor to believe that all of a university's resources are freely available for use, simply because they already exist and have been paid for by the university. In particular, an institution (or even a faculty or department) may run various operational units as cost centers, and each of these cost centers may charge for services and resources. For example, if a supervisor determines that a postgraduate research program will require the use of electron microscopy facilities, medical imaging facilities or technical workshops, then the supervisor needs to determine how the use of these facilities is costed and charged.

Sometimes, institutions impose corporate charges even on the provision of basic facilities. For example, an institution may charge for items including:

- Recruitment costs if a graduate student position is to be advertised.
- Information technology (IT) and support.
- Office space, printing and stationery costs for the research student.
- Library support.
- Purchase of journals from other institutions.
- Cost of publishing findings in publication-fee-based journals.

A supervisor needs to determine what these charges are and how they will be funded in the context of any proposed research program.

In addition to these charges, it may be necessary for a research student to attend national/international conferences or other functions, and so travel costs need to be considered Even if a research student is asked to travel from the university to local destinations (e.g., from a university campus to a teaching hospital), then it may be necessary to provide reimbursements for mileage/car expenses or cab fares.

One useful approach to dealing with these issues is to prepare a table, such as the one shown in Table 6.1, in order to ensure that all the key costing/resourcing elements are included. This can then be discussed with departmental heads or other senior colleagues to ensure that it comprehensively covers the requirements of the project.

In addition to funding, there are other issues that need to be considered in planning a postgraduate research program. One of these is consideration of the ethics requirements of the project, in the event that it involves human or animal experimentation. A research supervisor should speak with experts in the field, or with those charged with ethics approvals in order to get an insight into whether the project is viable from an ethics perspective – *before* commencing the project – not after a project has commenced, when it becomes apparent that ethics obstacles may be insurmountable.

Additionally, in defining the project, and in the context of ethics approvals, a supervisor may need to determine if a research student is required to obtain certification from an accredited training provider in order to conduct the research. If this is the case, then the cost of ethics training/certification may also need to be included in the project costing/funding.

Table 6.1 Example of costing for postgraduate research project

Funding Requirement	Specific Items	Cost	Source
Project Specific	• Scholarship • Equipment Item 1 • Equipment Item 2 • Technical Support	• $90,000 • $8,000 • $12,000 • $23,000	University Seed Funding Grant
Project Support	• 120 Hours Use of Electron Microscopy Facility	• $12,000	Faculty Funds
Corporate Costs	• Recruitment • IT • Office Space • Stationery and IT Sundries • Library • Publications	• $18,000	Department Funds
Travel Costs	• Conference • Local Travel	• $6,000 • $2,000	Departmental Travel Budget
Total Costs		$171,000	

There are numerous other issues which the potential supervisor will also need to consider. Some of these may be basic, practical ones that can halt the entire program if not considered from the outset. A good example is the availability of specialist equipment or technical support. If a unique facility/resource is required for the research student, and the facility is unavailable (e.g., because it is being renovated, reconstructed or fully-booked out to other projects), then this could completely negate any chance of the student completing his/her research program.

To summarize – a relatively common mistake that novice supervisors make, in the enthusiasm to start their project, is not to clearly identify the costs and funding sources for the research. Sometimes, there is an assumption that the money will simply be provided by the university and often this is not the case. The starting point, therefore, is to ensure that the project:

- Is clearly defined.
- All the costs are identified.
- The funding sources and resources are identified and preferably locked-in.

6.3 Scholarship Funding

6.3.1 General Issues

Students undertaking postgraduate research degrees are adults, many of whom will have adult responsibilities, including families, rent, mortgages, car expenses, and so on. Those that don't have families when they commence their research may choose to do so at some point during the research program. In general, therefore, keeping in mind the age of those undertaking research, it is unreasonable to expect postgraduates to simply support themselves financially for several years or more via external means.

For these reasons, in many universities around the world, postgraduate research programs are linked either directly or indirectly to scholarship funding – particularly for studies undertaken at a Doctoral level. Some universities avoid taking in research candidates without scholarship funding, because the likelihood of successful completion is diminished when postgraduates need to earn a supporting income from external sources.

Postgraduate research scholarship funding can arise from numerous sources:

- In some countries, scholarship funding is awarded by national/regional governments to institutions based upon their overall research performance from preceding years.
- In other countries, the funding is provided as part of a university's normal, recurrent funding.
- Some institutions use income from student fees or endowments to fund research scholarships, and others use commercial income.
- When universities enter into collaborative research programs with partner organizations (e.g., industry) it is also common for the collaborative project budget to include funding for postgraduate research students to work on nominated projects.

In keeping with other research funding schemes, postgraduate scholarships tend to be linked directly to either personal or institutional academic achievement. In the case of individuals applying for scholarships, this often occurs through a formula that recognizes various subject results realized during undergraduate studies. Additional consideration may be given to academic awards; published research papers and conference proceedings, or other postgraduate coursework studies.

Each national/regional and internal university scholarship system has its own characteristics and weightings but the underlying truth of the matter is that there are far fewer scholarships than there are applicants.

Formula-based scholarship awards have the advantage of being straightforward to administer and, for students all emanating from the same undergraduate programs, provide a reasonably objective decision-making process. The downside is that the process is very mechanistic and often disadvantages international students who come from different educational systems which:

- Award marks on a different basis.
- Use different benchmarks for academic grades or degree rankings – for example, some institutions may only award a single first-class honors degree each year – other institutions may award many.

There is also the commonly raised issue that the scholarship system rewards those with strengths in rote learning, who can achieve high undergraduate grades, even when their capacity to innovate or act as free-thinkers – important characteristics for a researcher – is limited.

In some cases, university faculties, departments, research institutes or centers award their own scholarships, and these may provide a greater degree of latitude in the selection process.

The end result, however, is that only a small proportion of potential postgraduate research applicants qualify for research scholarship funding. This means that a research supervisor either has to accept those who have qualified for scholarships, or else find a separate funding source for a student who does not have scholarship funding, but which the supervisor feels may have particularly useful attributes for the intended research program.

6.3.2 Scholarships Arising from Collaborative Research Programs

When postgraduate research funding is part of a larger collaborative program with an external partner (i.e., commercial or external research partner organization), then numerous additional challenges arise:

- The external partner may insist on being an arbiter (or even final arbiter) on the appointment of a research candidate for scholarship funding.
- The preferences of the external partner may clash with university research candidature entry requirements or the preferences of the supervisor.
- The larger research program may have a contractually binding intellectual property (IP) distribution which is imposed upon the postgraduate student, should he/she choose to accept the scholarship.
- The larger research program may have contractual confidentiality requirements which are imposed upon both the supervisor and the research student – and which will restrict the ability to publish outcomes.

None of these are straightforward problems to resolve, and each requires – on the part of the supervisor – a level of maturity and capacity to negotiate when disagreements/disputes arise. It also requires a level of strength on the part of the supervisor to ensure that he/she does not negotiate away basic requirements of the postgraduate research program that would jeopardize its completion. For example, a collaboration agreement may have confidentiality requirements which are such that a student would not be permitted to submit a dissertation to external examiners – such a constraint would be untenable from the university's perspective.

A postgraduate research student, accepting an offer of a scholarship which has conditions that are not normally associated with a research degree (i.e., confidentiality, intellectual property distribution, etc.), should be cautioned to seek independent, professional (legal) advice before signing an acceptance. If, for example, a research student has aspirations of creating a start-up company at the completion of the research program, then an IP agreement signed at the beginning could preclude the use of many of the elements and knowledge developed during the course of that program. Importantly, a research student needs to be cautioned that a university's legal representatives, who may negotiate the terms of a postgraduate scholarship with an external partner, are there to negotiate in the interests of the university – and not necessarily the research student. The student therefore needs to take advice from independent legal sources.

6.3.3 Other Issues

In some nations and institutions, scholarship funding for postgraduate research has a special legal status that is distinctly different from a conventional salary. For example, some countries may deem that university research scholarships are exempt from taxation. More generally, however, national, regional or institutional guidelines will distinguish between the payment of a research scholarship and the payment of an employee salary. The distinctions between scholarships and salaries generally lead to restrictions in the sorts of activities that a research student can undertake while on a scholarship.

A research supervisor therefore needs to be fully aware of the restrictions that apply to the management (i.e., supervision) of a research student relative to the management of other academic staff. Typical restrictions may include the following:

- A student on a research scholarship may not be permitted to take part in day-to-day academic work (e.g., teaching, grading of papers, etc.) unless an additional contract payment is made.
- A research supervisor may not have the authority to compel a research student to attend the institution during particular hours (e.g., 9am–5pm).

- A research supervisor may not have the authority to compel a research student to undertake any extraneous university activities (e.g., assisting in experimentation or equipment set up) beyond what is specifically related to the postgraduate research program.
- A research student may have no intrinsic entitlement to annual leave/vacation or paid sick-leave entitlements.

6.4 Student Recruitment

6.4.1 General Issues

There are a wide range of pathways which may bring a potential research student and supervisor together for the purposes of a postgraduate research program. These include:

- A research student, who is familiar with the supervisor and his/her work, elects to work with that supervisor after having been awarded scholarship funding.
- An international student, who has never physically met the supervisor, has notionally agreed to work with him/her in the event that they are awarded scholarship funding.
- A research supervisor, using funding from a collaborative research program, is part of a decision-making committee that interviews potential research students and makes a selection as would be the case for a research employee.
- A research supervisor is awarded a competitive research grant, which includes scholarship funding, to allocate, at his/her discretion, to an eligible student, provided that institutional selection criteria are met.

There are numerous other possibilities, but even the few listed above suggest that the bringing together of supervisor and student can occur:

- At the initiative of the student, and based upon his/her ability to attract competitive scholarship funding in his/her own right.
- At the initiative of the supervisor, based upon funding that he/she has derived from some research funding mechanism.

The former possibility is the classical, academic connection that takes place as a result of negotiations and camaraderie between students and supervisors. The latter is more akin to a normal professional recruitment and appointment – except that, instead of salary, a student is paid a stipend.

It is unlikely that a university, operating in a modern, professional environment, will simply allow a research supervisor to select a research student at whim and make an appointment without justification. It is far more likely that a supervisor will need to go through a formal process of recruitment and appointment, which fits in with a university's basic personnel requirements, including fairness, non-discrimination, etc. In some institutions it may also be the case that an individual supervisor cannot make a scholarship award at his/her own individual discretion, and that the process needs to be carried out through an interview panel, constituted according to university procedures.

6.4.2 Selecting the Appropriate Candidate for Postgraduate Research

In situations where the selection of a research student has become a matter for the supervisor (either as an individual or as part of a selection panel), a range of subjective issues will inevitably arise, and some familiarity with what these may be is useful. In the case of a novice research supervisor, the selection of the first postgraduate research candidate may be the first time the supervisor has ever had to appoint an individual to a professional role.

An experienced research academic will be only too aware that any human recruitment process is intrinsically subject to errors and inaccuracies, In the long-run, even for experienced individuals, the professional judgment used to select candidates may be wrong as often as it is right. For the novice researcher, however, there is no long-run, and a mistake in recruitment is tantamount to 100% failure on a single project. The stakes therefore appear to be much higher, so there is a greater impetus to try and *get it right* The reality is that in a mature recruitment process, there is no *right* or *wrong* – there are just decisions with shades of gray.

For numerous reasons, including the inability to simply discard a research student at whim, the decision that a supervisor makes is one that he/she will need to live with (and make the best of) for several years.

The basic logic associated with the selection of candidates is relatively straightforward. In order to achieve good research outcomes, the supervisor needs to recruit a *high-caliber student*, wherein a *high-caliber student* is defined as one who will potentially outshine the supervisor. If supervisors have inhibitions (i.e., lack of self-confidence) about seeking students who are potentially better than they are, then perhaps they should ponder on this long and hard. After all, a research student who requires constant instruction from the supervisor is not really a researcher but a personal assistant.

The supervisor and the student need to be able to challenge each other (and each other's views) intellectually so that both can grow, and so that the whole can be greater than the sum of its parts.

In the long-term, a supervisor will not be judged solely on his/her own achievements but the achievements of those that he has supervised. A research student who ultimately goes on to achieve greatness in research/academia; industry; business; government, etc. can become a testament to the supervisor's achievements as well.

It also needs to be noted at this point that if the selection process goes awry, and the postgraduate research candidate does not live up to initial expectations, then it will be a testament to the supervisor's skills if he/she can still lead the student to achieve good research outcomes, and to become an outstanding professional.

With these points in mind, there is obviously a need for the supervisor and the potential student to meet face to face – either at a formal interview session or an informal chat. In some cases, because potential students are international, a video conference may need to suffice. As a general rule, it is not a good idea for a supervisor to make a selection without any form of discussion with the candidate – and based solely upon a written curriculum vitae. At the end of the selection process, the two parties will need to have a professional working relationship for several years, and written correspondence does not provide a complete picture of an individual's capabilities. In particular, written applications tend to compress or suppress intricate nuances that may be critical to the selection process. And, needless to say, without additional evidence or verification, there are no guarantees that a presented curriculum vitae is even written by a potential candidate.

There are numerous issues to consider when conducting a discussion with a potential candidate about a postgraduate research program, and few of them are clear-cut. Table 6.2 lists some of the potential candidate traits that may arise during a selection process, and the sorts of issues that a supervisor should consider before making a selection.

Finally, there is the need to consider the intellectual match between the project and the research candidate. If a research project is intellectually below the stimulation level required for a high-caliber candidate, then that candidate may get bored with the project and leave mid-way through. If the project is above a candidate's intellectual capacity then clearly the candidate will not be able to achieve worthwhile outcomes. In general, however, a high-caliber candidate, given sufficient motivation and leadership by the supervisor, will

Table 6.2 Candidate attributes that supervisors need to consider

Candidate Attributes	Issues to Consider
Academic Grades – High	• Are they indicative of a high level of intellectual capacity or just a propensity for rote learning and structured problem-solving environment?
Academic Grades – Low	• Are they indicative of low intellectual capacity or a creative inability to conform to structured undergraduate learning – is the candidate intellectually "above" the undergraduate learning process?
Loud/Assertive Personality	• Does this indicate arrogance and misguided sense of self worth or a capacity to cut through problems when necessary?
Quiet/Subdued Personality	• Will it be possible to work with the candidate? • Will the candidate be able to work with colleagues, technical and laboratory staff? • Will the candidate be able to get things done when assertiveness is required with other people? • Does the subdued personality indicate modesty and an individual who considers things before acting rashly?
Flattering/Gushing	• Does the supervisor want someone who just feeds his/her ego or someone that will challenge every idea in order to get better research outcomes?
Combative Personality	• Will the candidate be disruptive to the research group and disturb the work of others?
	• Is the supervisor an individual who relishes intellectual arguments or avoids confrontation?
Career Planner/Projector	• Does the candidate have an oversimplified view of his/her professional trajectory?
Over-Ambition	• Does the candidate have the perseverance required for a project lasting several years or will he/she just flit from one short-term opportunity to another?
Lack of Ambition	• Does the candidate lack ambition because of laziness and lack of self-motivation or because they require a significant challenge? • Can the supervisor provide the sort of challenge that will spark ambition?
Attention Span	• Does the candidate get easily distracted by techno-gimmicks (e.g., smart phones)? • Does the candidate require constant stimulation to retain interest? • Can the candidate avoid distractions and focus on a rigorous, disciplined research pathway?

take any project and drive it in his/her direction – perhaps surpassing greatly the initial expectations of what starts out as a mundane piece of research.

6.5 Submission of Candidature Forms

6.5.1 General

The university processes associated with the submission of a postgraduate research candidature form vary greatly from institution to institution, and country to country. At its most basic level, a candidature form is a contract that clearly enunciates the program of investigation that is to be undertaken by a postgraduate research student, in order to achieve a higher degree by research. Essentially then, it is a formal, three-way agreement between the university, the supervisor and the student. It therefore needs to be taken seriously by all parties.

Importantly, from the perspective of the supervisor, the candidature proposal should not be dismissed merely as a piece of annoying bureaucracy that gets in the way of creative research. It is the reference framework for the student, supervisor and university for the duration of the research program. To treat it as less than such is to do an injustice to the research student.

Consider particularly a situation where one of the parties to the candidature agreement is unable to deliver its side of the contract, and the research student is subsequently unable to complete his/her higher degree outcome. In the modern world, there is a real prospect that litigation will arise.

For these reasons, a candidature form generally contains a number of academic, procedural and managerial aspects, specifically detailing issues such as:

- The title and program of research to be undertaken, including some notional methodology.
- The expected outcomes of the research.
- The capacity (skill-set) of the supervisor/s to oversee such a program of research.
- The capacity (academic record) of the research candidate to undertake the program of research.
- The resources to be provided to the candidate by the university in order to complete the program of investigation.
- The ethics requirements of the investigation, where applicable.

Implicit in this three-way contract is an understanding that if each of the three parties delivers on its commitments then, following examination, the research candidate has a high likelihood of being awarded a higher degree by the institution.

6.5.2 Candidature Processes

The processes for higher degree candidature vary considerably from country to country and institution to institution. For example,

- In some institutions a potential research student needs to work closely with a supervisor to prepare the candidature proposal in a joint effort.
- In other institutions (particularly where international students are involved), a potential research student may prepare his/her own candidature form independently of the supervisor, and only meet with the supervisor to discuss it after the proposal has been considered and approved by other committees.
- In some universities, potential students need to submit candidature forms in conjunction with scholarship applications.
- In other institutions the scholarship process and the candidature process are completely independent issues.
- In US universities, a thesis committee is often used, with the supervisor chairing that committee. It is also common for a candidate to have a *qualifying exam*, often oral, after a presentation to a qualifying committee. The subject matter is often an extensive review of background literature. A student who passes the qualifier then advances to candidature for a higher degree such as a Doctorate, and is overseen by the thesis committee.

With these points in mind, it becomes apparent that there are numerous potential entry points to a postgraduate degree program, depending upon both the country and institution in which it is conducted.

For these reasons, and because a supervisor is one of the key parties to any candidature agreement, he/she clearly needs to be on top of all relevant university processes that are in place to deal with the preparation and submission of an application. Ideally, the supervisor needs to scrutinize closely any research proposals submitted by potential research candidates. A supervisor needs to be satisfied that any submitted candidature proposals:

- Present a reasonable opportunity for the research student to achieve his/her objectives within the allotted research time-frame.
- Are research activities which are reasonably within the supervisory capacity of the supervisor.

- Can be accommodated in terms of laboratory space, equipment and other resources.
- Where applicable, that there is a high likelihood of the research (and the research student) being given appropriate ethics approval and/or certification (accreditation).

There are, however, numerous logistical problems associated with the candidature process. In a traditional postgraduate research process, a final-year undergraduate student may simply approach an academic/researcher and the two may mutually agree to participate in a specific postgraduate research project.

However, this does not accord with many modern practices. For example, in some large universities, there are thousands (sometimes tens of thousands, when including international students) of potential postgraduate research applicants, who need to be matched with potential supervisors. In many cases, the students and the supervisors have never met – sometimes the two are located in different countries. A hosting university may also have separate operational units and committees to deal with postgraduate research and scholarship applications. So, things are considerably more complex than just having a potential supervisor and student agree to work together.

In the context of growing internationalization of the university sector, it is increasingly common for an international postgraduate applicant to identify an academic within a university (from a website), then submit a candidature form, independently of the supervisor, to the university. Many universities have online, automated candidature application processes – some with automated administrative workflow at the back-end. Inevitably, given the volumes of applications that may arise, it may be that some research projects proposed by students are completely impractical in the context of resources available at the university, or simply outside the supervisor's specific field knowledge.

The point here is that there need to be checks and balances within the university system to prevent mismatches from happening. However, the reality is that in large institutions, having numerous operational units; committees with overlapping jurisdiction; offering thousands of scholarships to a shortlist derived from tens of thousands of applicants, and matching these with thousands of supervisors, anomalies will inevitably arise. Supervisors need to be able to identify and manage these anomalies, mismatches and other errors in a professional and courteous manner which reflects well upon the university.

Regardless of the processes that are in place, the buck ultimately stops with the supervisor and, regardless of the point at which the candidature application form finally materializes, it is the supervisor's responsibility to ensure that the contents are entirely accurate and feasible – or that the application is immediately altered to something which represents the mutual agreement of the three major parties to it.

6.5.3 Ethics Considerations

In some universities, the submission of a research candidature proposal, in which the proposed program of research includes either human or animal related experimentation, including even mundane matters, such as surveys, it may be necessary for the research candidate to obtain ethics approval. The processes related to ethics approval will be specific to each institution. However, they are often complex and, in some universities, ethics approval may also require that the research candidate achieves certification or accreditation in some ethics related training program.

The research supervisor has a responsibility to ensure that he/she is fully apprised of the ethics requirements of the university and the proposed research project, before agreeing to undertake any supervision that will involve ethics approval.

In particular, a supervisor needs to factor ethics related issues into the timeframe of a postgraduate project and then determine (estimate) whether or not that project can still be completed within the allotted candidature time.

6.5.4 Timeline Considerations

Research students who are local to the university may have some latitude in the total number of months that they can dedicate to a postgraduate program. The same may not be true for international research students. There may be constraints on visas, work permits, etc. that restrict the total number of months that international students can devote to their candidature.

It is incumbent upon the supervisor to consider carefully and realistically (i.e., without optimistic assumptions) the actual time that a candidate will require in order to complete his/her program of research. If a supervisor genuinely feels that a candidate will be unable to complete the work outlined in his/her candidature proposal, and it is clear that the candidate does not have latitude in extending candidature, then the supervisor should act early to initiate revision of the proposal.

A supervisor who allows a time or resource infeasible proposal to go through the candidature application process, in the hope that good fortune might intervene during the candidature is, in the best interpretation, naive and, in the worst interpretation, acting unethically. A failure on the part of the supervisor to provide an early intervention may lead to a research candidature which is never completed, or which fails the examination process.

At all stages in the candidature application process a research supervisor needs to understand that he/she is dealing with matters of more significance than mere university bureaucracy. Applicants for research candidacy are often staking their dreams and career aspirations on the outcomes of a candidature/scholarship application. For some international applicants, particularly those emanating from undeveloped or developing countries, acceptance into a world-class university could not only change their futures but also those of their families and possibly surrounding regions. So, while the candidature process may be bureaucratic, one should never lose sight of the fact that the outcomes of that process and the selections made could significantly change people's lives.

7
Initiating the Formal Supervisory Process

7.1 Introduction

It is difficult to determine statistically what proportion of postgraduate research programs fail as a result of the supervisor and what proportion fail as a result of the student. Generally, neither supervisors nor students are eager to accept blame for an unsuccessful program. Of course there are also other extraneous, non-academic factors that may contribute to the unsuccessful termination of a project, including:

- Personal bereavement
- Illness
- Relocation
- Initiation of a start-up company
- Research student being head-hunted for a lucrative job prior to completion.

Notwithstanding non-academic factors, in an unsuccessful project, where things have gone awry, the student is likely to blame the supervisor and the supervisor is likely to blame the student. Suffice to say that, because it is the supervisor's responsibility to make a postgraduate research program work then, if it doesn't, the supervisor generally has to accept responsibility for the failure.

There are naturally a few instances where a research student needs to wear responsibility for the failure of a postgraduate program. These include cases where the research student:

- Has been lazy and/or unproductive, despite the best efforts of the supervisor.
- Does not, in practice, have the intellect, academic skills or capacity for independent thought and work required to complete a program.

These sorts of occurrences tend to be the exception, however. In general, as long as a student has successfully completed undergraduate studies at a high level, and is genuinely contributing work effort to the process, it is the supervisor's responsibility to make sure that that student reaches the required standard.

The place where a supervisor's failings are most likely to manifest themselves are in the early stages of the research program. In the initial phase, a supervisor's role is far more onerous than just issuing instructions and requirements, on the assumption that the student is able to carry them out. The supervisor has a responsibility to initiate the program by:

- Trying to understand the student's approach to learning.
- Identifying student strengths and limitations.
- Taking the time to develop a suitable supervisory approach which will build on student strengths and specifically address any limitations.

This must be a *closed-loop* process. In the early stages of the research program this process may require the supervisor to:

- Set time-limited tasks for the student in order to provide a vehicle for capability assessment.
- Assess the capacity of the student to deliver results in the time-constrained period.
- Vary the time-frame for the next task and assessment, based upon the student's performance on the current task.

This initial activity is clearly also an iterative process, and can be a particularly time-consuming one for the supervisor. However, it is at this point where the supervisor can either set the foundations for a successful program, by putting in the initial hard work, or just let things slide – and accept that, at some point down the track, the program may ultimately become irretrievable because of compounding failures that have not been addressed in time.

The corollary of the above discussions is that it is imperative for the supervisor to inject maximum effort at the start of the program. This doesn't mean that the supervisor needs to become a *nanny* to the student – perhaps, it may mean the exact opposite and require the supervisor to challenge the student and allow him/her to fail and learn from relevant exercises.

With these points in mind, exactly when is the start of a postgraduate research program?

Although this may appear to be a question with a relatively obvious answer, in practice, it is sometimes difficult to identify a specific starting point for a

7.1 Introduction

postgraduate research program. The actual starting point for some programs may be a casual conversation held with a final-year undergraduate student after a lecture – and months before the actual postgraduate candidature process commences. For other students, the starting point for research may take place after the formal submission and acceptance of a candidature form.

Herein, we are specifically concerned with what takes place after the formal acceptance of candidature, and when a research student is formally recognized as a postgraduate student entity within the university.

As a general rule, if a relationship between a supervisor and student starts badly, then it is unlikely to improve during the course of the research program, so the supervisor is responsible for ensuring that it does not start badly. There are two broad aspects to initiating the research program:

- Administrative/managerial – ensuring that correct university processes will be followed during the program.
- Academic – formally initiating a systematic program of investigation in the field of interest.

Academics and researchers often underestimate the importance of the administrative/bureaucratic side of research supervision. The reality is that the administrative/bureaucratic processes pertaining to research supervision can be as complex as the research being undertaken. It would therefore be both naive and arrogant for a supervisor to just brush these issues aside as irrelevant bureaucratic trivia, simply because their innate complexities do not fall into the supervisor's chosen field of research interest.

Supervisors need to develop sufficient maturity to understand that, while they may have an underlying passion for knowledge and learning in a specific area, the reality is that they work in a large, complex environment which has:

- National/regional legal operating frameworks.
- Institutional operating/procedural frameworks, based upon external legal requirements and internal structural requirements.
- Potential health and safety threats that may impinge upon the research student.
- Financial and business constraints.
- Complex academic structures for governance and oversight of research activities.
- Complex internal and external reporting requirements (e.g., for research outputs).

These cannot simply be ignored, and dealing with them professionally is an integral part of the supervision process.

In this chapter we look at some of these issues as well as the more fundamental ones of the research program itself.

7.2 General Induction for Research Students

Universities can be very large organizations, and most large organizations do not simply allow individuals to come aboard without some form of induction. Postgraduate research students may not technically be staff of the university but their role is such that they are more akin to staff than to students. Research students have much greater levels of access to university resources than undergraduates, and far less direct supervision when in a laboratory.

A university/faculty/department/center/group may therefore deem it to be appropriate for every research student to be formally inducted into the complexities of the university environment.

Some universities regularly run induction programs specifically tailored for postgraduate students, wherein new cohorts can interact with one another, and with more experienced postgraduates – or academic staff. There are numerous issues that need to be covered through induction. These may include:

- General university rules and procedures pertaining to postgraduate studies.
- Rules and regulations pertaining to scholarships.
- Rules and regulations pertaining to ethics.
- Discussions on postgraduate assessment processes within the university.
- Issues relating to acceptable conduct within the university – specifically in relation to areas such as discrimination, harassment, bullying, etc.
- Workshops on oral presentations, preparation of research papers, dissertations, etc.

A research supervisor needs to be aware of the specific induction training that is available at university, faculty or departmental level. If there are any gaps, then the supervisor has an obligation to ensure that they are filled appropriately.

7.3 Occupational Health/Safety

A supervisor has a moral obligation and, in many jurisdictions, a formal legal obligation to ensure that individuals operating under his/her management do so in a safe manner, and without subjecting themselves – or other co-workers – to short or long-term health risks.

Each research group, center, department, institute, faculty and university has its own unique set of operating conditions, and clearly it is not possible to have a generic set of rules covering every eventuality. However, each supervisor should be in a position to undertake a professional examination of his/her environment to determine what specific approaches need to be adopted in order to ensure the safety of the research student. It is a serious error of judgment to assume that a research student will naturally be able to determine these for himself/herself.

A supervisor needs to understand that a research student, in his/her enthusiasm to complete a task, may take risks which – from a student perspective – appear reasonable but which, to a more mature professional, are clearly dangerous. A supervisor also has the responsibility of interpreting departmental/faculty and university regulations in regard to health and safety, and seeing how they impinge upon his/her area of operation. Consider Example 7.1.

Example 7.1
A university has an electronics engineering department, which has established operating procedures and rules for working with electronic (i.e., low power) equipment. A supervisor comes into the department and decides to conduct a postgraduate research program in high-power electronics. Prior to commencing the postgraduate program, the supervisor may need to:

- *Liaise with departmental staff in order to establish formal (documented) operating procedures for the high-power equipment experimentation.*
- *Establish safety rules which are additional to any existing departmental rules (e.g., forbidding students from working in a high power laboratory without the presence of another student or staff member).*
- *Put in place additional training regimes specific to the equipment (e.g., provide CPR training to all staff and students working in the high-power electronics laboratory).*

Academics can sometimes dismiss this sort of attention to safety detail as annoying bureaucracy but the reality is that it can be a critical element of the safety of any research program.

Universities are likely to already have template guides for the development of procedures relating to health and safety. If not, then a supervisor may need to develop his/her own as they specifically pertain to the local environment.

The stakes as they pertain to health and safety of research students are extraordinarily high. A failure can manifest itself in death or serious,

134 *Initiating the Formal Supervisory Process*

permanent injuries, and the research supervisor is likely to bear responsibility for these in a moral and/or legal sense. In some jurisdictions, a failure to provide a safe working environment can lead to criminal sanctions, over and above any punitive measures imposed by the university itself.

From a moral and legal perspective, therefore, there is no element in a research program more important than the basic health and safety of those working in the research environment. Supervisors have a responsibility to ensure that their students understand and accept this.

7.4 Initial Meeting with the Research Student

7.4.1 General

Once a supervisor is confident that he/she is on top of the administrative/bureaucratic aspects of a candidate's supervision, there needs to be an initial meeting between the candidate and the supervisor to formally commence the academic/research aspects of the process.

The nature of the first meeting ultimately depends upon the relationship that a supervisor has with a student. Some supervisors may have known their research students for years, during their undergraduate degree programs. Some may have already supervised the research students for undergraduate thesis projects. For many supervisors, however, the initial meeting with the student will be the first time that the two will discuss the project in a formal and clearly defined way as part of a structured research program.

The supervisor should allow a significant amount of time for the first meeting. Ideally, it should be open-ended and stretch for as long as is required for both parties to leave the meeting, comfortable with the arrangements that have been agreed. Depending on the specific nature of the research program, and the capabilities of the student, the initial meeting could conceivably extend to several hours duration.

Some supervisors make the mistake of using their initial meeting with a research student to *lay down the law* and assert their authority. Needless to say, this not only reflects a lack of understanding of the practical, authoritative scope of a supervisor, but can also provide a fast-track to alienating the student and setting up the basis for a combative relationship.

A supervisor can't assume or demand respect from a research student – even if the supervisor is an eminent authority in his/her field. A research student is likely to be a highly intelligent person in his/her own right, and the fact that they have not yet had the opportunity to develop eminence is no excuse for a condescending attitude on the part of the supervisor. The research student's respect is something that needs to be earned – and earning it is a long-term process.

7.4.2 Understanding the Research Student's Perspective

The initial meeting between the supervisor and student should provide an opportunity for the supervisor to find out as much as possible about the student, in order to develop an optimal research learning approach. There is little value in discussing the academic aspects of a research project until the supervisor gets a solid understanding of the student's background, vis a vis:

- General cultural issues.
- Schooling and university studies.
- Learning preferences during undergraduate study.
- Specific subjects that the student found of interest.
- School or university teaching/lecturing staff that inspired the student.
- School or university teaching/lecturing staff that the student didn't like.
- Social activities/sports interests/hobbies.
- Family commitments and dependencies.
- Lifestyle issues – is a research student living at home, or has he/she moved away for the first time?
- International student issues – how is the student settling in (coping) with the change of country and culture?
- Extra-curricular business, employment or volunteering interests.

Once a supervisor has established the student's background, there needs to be some discussion as to the student's future directions. A cardinal mistake in determining this is to ask the cliché question,

"Where would you like to see your career in five years time?"

This is only going to corner the student into providing a cliché answer that might please the supervisor,

"I'd like to become a professional researcher and ultimately a tenured academic."

A better approach is for the supervisor to start by explaining the advantages and disadvantages of an academic career – and then talk about other potential options including:

- Business
- Industry
- Government R&D
- Start-up companies.

The discussion should also include the sorts of preparatory work that the student will need to undertake (in addition to his/her basic research) in order to achieve these sorts of objectives. This should provide ample opportunities for the research student to interject and contribute – and for the supervisor to genuinely determine the relative levels of enthusiasm that the student has for each option.

All these discussions need to be meandering in their approach in order for the supervisor to gently prompt the student to express his/her genuine preferences and explore options, rather than just have the student enunciate answers that he/she believes are required responses.

A supervisor needs to be a good listener during this phase of the meeting – listening for spontaneous bursts of enthusiasm, disdain or revulsion whenever various topics are broached. These bursts are the ones which contain the genuine, unfiltered information that the supervisor will need in order to re-orient his/her own thinking about how to approach the project. The old adage that humans are equipped with two ears and one mouth, which should be utilized in those proportions, is particularly relevant here.

7.4.3 Negotiating a Mutually Acceptable Supervisory Approach

A supervisor's role is not simply to provide the pathway of least resistance for a student to achieve his/her ends but, more importantly, to develop a considered program of learning that will challenge a research student and enable him/her to grow professionally. This may require the supervisor to take students outside of their comfort zones, extend themselves, and to do things that they might not otherwise do of their own volition.

For these reasons, a supervisor needs to have some understanding of the sorts of approaches that a student is comfortable with, but these should not limit considerations on the learning program. For example, some research students may be:

- Introverted and uncomfortable with interacting with other people.
- Poor/nervous public speakers.
- Intently focused upon the theoretical technicalities of their research to the exclusion of practical or commercial realities.

7.4 Initial Meeting with the Research Student

- Poor at writing and have limited skills in the English language.
- Self-serving career-climbers who will overstate achievements; take shortcuts, or misrepresent information in a positive light in order to get ahead.
- Obsessive publishers who want to turn every minor experiment into a research paper.

These are the sorts of issues that require the mature judgment of a professional and they should not be taken lightly. If, for example, a student is highly introverted, a research supervisor needs to consider:

- Is the student likely to benefit from being forced into a situation where he/she needs to interact with others on a regular basis?
- Is it reasonable to coerce a timid student into a position where he/she has to interact and thereby cause them considerable psychological stress?
- Is the introverted student likely to become extroverted as a result of any artificial activities/environments created by the supervisor?

In other words, a supervisor needs to make a subjective decision as to whether he/she can act as a positive agent of change on the part of the research student. And, at the same time, it is important that the supervisor does not attempt to take on the role of an amateur psychologist – if there are concerns, then the student should be referred for professional guidance elsewhere.

Nevertheless, as a result of extensive discussions with the research student, a supervisor needs to notionally determine a set of strengths and limitations, and then work out how (or if) the limitations can be reduced or turned into strengths.

In addition to the basic strengths and limitations of character that a student may have, a supervisor also needs to consider the academic capacity of the research student – and not simply in the context of academic grades from undergraduate programs. Consider Examples 7.2 and 7.3, which occur commonly in supervision.

The key point arising from these examples, and numerous other possible scenarios, is that the supervisor should not let things slide. A supervisor needs to be on top of the process at all times. Specifically:

- Identify student capabilities.
- Identify a strategy for managing according to capabilities.
- Ensure that actual mechanisms are in place to manage the student (either visibly or subtly).

Example 7.2 – Dealing with a Rote-Learner

Situation:
This is the case of a student who has achieved high academic grades at undergraduate level but who originates from a country where undergraduate learning is highly rote-based and procedural. It is naive for a supervisor to believe that if the student is, "thrown in at the deep end", that student will magically be capable of high levels of independent research.

Possible Supervisory Approach:
Such a student needs to be transitioned by the supervisor from the rote learning environment to independent research. The transitioning process may take months – and it will require considerable effort on the part of the supervisor. The supervisor may initially need to set very short-term objectives/milestones for the student (e.g., weekly) and then see how he/she handles these. If a student is unable to work to weekly milestones, then the timeframe needs to be shortened to daily objectives – or even half-daily objectives – until positive outcomes result and the timeframes can be progressively lengthened to something meaningful in a research context.

Considerations:
A student is not a failure, or incapable of performing good research, simply because he/she has been hamstrung by an inadequate learning process at undergraduate level. A research supervisor has the capacity to resolve this problem.

Example 7.3 – Dealing With a Strong-Willed High Achiever

Situation:
A high-caliber student, with no significant research experience, wants unfettered freedom to do the postgraduate research as he/she please and to largely act independently of the supervisor.

Possible Supervisory Approach:
If a research student is extremely capable, and there is sufficient latitude in the description of the research program, a supervisor may elect to give that student some headway in decision-making. Rather than scheduling formal meetings on a short-schedule, the supervisor may instead elect to simply visit the research student on a weekly basis for informal chats in order to provide a less visible form of oversight.

Considerations:
A highly capable research student may still have severe limitations in terms of research process and rigor. An enthusiastic student may meander in a broad range of unproductive directions, using up valuable time in the postgraduate program. A supervisor still needs to be a supervisor – and create more subtle means of ensuring rigor and direction.

7.4.4 Establishment of Formal and Informal Meeting Mechanisms

It is neither professional nor efficient for a supervisor to assume that he/she can simply work with a research student in an unstructured, friendly, collegial (mentor-like) manner. This is only one part of a supervisor's role. The supervisor has a responsibility to deliver outcomes for both the student and the institution, and some of these outcomes are dependent upon the establishment of a formal research learning structure. One of the elements of a learning structure in postgraduate research is a series of formally scheduled meetings at which:

- Progress is discussed in an open and frank manner.
- The supervisor and student exchange ideas/knowledge.
- Problems/impediments to the research program are discussed (e.g., resource problems, lack of access to laboratory staff, etc.).
- Student limitations are assessed and measures enacted to attempt to address these.
- Realistic goals/targets are established to be discussed at the next meeting.

The frequency, duration and timing of the meetings is a matter for the supervisor to determine, based upon:

- The student's ability to deliver on agreed outcomes.
- The student's capacity to work independently.
- The specific needs a student and supervisor have in relation to the research program (e.g., need to share updated datasets with other researchers in the group).

Clearly it is important that meetings do not become a time wasting exercise – where the supervisor and student merely go through the motions in order to check a box saying that they have had a meeting. If a supervisor observes that a meeting has been unproductive, then he/she has a responsibility to determine the root cause and ensure that it is addressed before the next meeting.

In the initiation phase of the research program, the research supervisor needs to work on the assumption that a student will require detailed guidance and support – particularly in relation to any initial administrative or resource access issues – so meetings need to be more frequent. As the student finds his/her legs, and becomes less dependent upon the supervisor, the length of time between formal meetings can be increased.

Each supervisor should be able to prepare a template – even if only as a mental picture – as to the nature, duration and time-lapse between meetings. Table 7.1 provides a sample.

Table 7.1 Sample template for meetings between supervisor and research student

Meeting	Duration	Subjects	Time-Lapse to Next Meeting
1	3 Hours	• Description of the institution, research group, etc. • Understanding student's background • Detailed discussion of research project • Discussion of initial tasks to be completed • Discussion of access to resources/staff (e.g., IT, laboratories, technical staff) • Introduction to other group members	1 Day
2	1 Hour	• Follow up meeting to ensure that resources have been provided and that student has settled in • Allocation of specific tasks to be completed by next meeting (e.g., literature review topics)	1 Week
3	2 Hours	• Review of work completed by student relative to previously set requirements • Discussion of issues/shortcomings • Allocation of new tasks	1 Week
4	2 Hours	• Review of work completed by student relative to previously set requirements • Discussion of issues/shortcomings • Allocation of new tasks	2 Weeks
5	2 Hours	• Review of work completed by student relative to previously set requirements • Discussion of issues/shortcomings • Allocation of new tasks	1 Month
6	2 Hours	• Review of work completed by student relative to previously set requirements • Discussion of issues/shortcomings • Allocation of new tasks	1 Month

7.4 Initial Meeting with the Research Student

Needless to say, the sorts of meetings described in Table 7.1 only constitute the formal meeting requirements. A supervisor will also need to meet informally with the research student on a more regular basis to ensure that the research student is:

- Comfortable with his/her activities and daily progress.
- Working satisfactorily in his/her environment and with colleagues.
- Generally well from a physical/mental point of view and not requiring additional support for personal issues.

7.4.5 Providing Constructive Feedback to Research Students

A particularly important aspect of postgraduate research supervision is the provision of ongoing, constructive feedback to research students in terms of:

- Specific, technical aspects of the work being undertaken.
- General research directions.
- Strengths and limitations of the work.
- Possible new directions to be explored.

None of these items should come as a surprise, even to a novice supervisor, but it is necessary to devote some effort to looking at the manner in which this sort of advice needs to be provided. Specifically, feedback advice needs to be:

- Timely.
- Relevant to the research student's needs and capabilities.
- Constructive.

Of all the complaints leveled at research supervisors, however, one of the most common relates to a failure to provide timely feedback.

The first priority of all academic and research staff in a university is in servicing the professional needs and welfare of students under their care. All other academic activities, regardless of their seeming importance, need to be rearranged accordingly around student priorities. Those academic and research staff that are unable to prioritize their commitments to ensure that students are at the top of their list, should consider whether they have made an appropriate career choice by working in academia and, more importantly, whether they are suited to the task of postgraduate research supervision.

It is unacceptable for research supervisors to rationalize a lack of timely feedback as being the result of other commitments or workload. Excuses for lack of timely feedback are neither of relevance nor interest to a student, who is fully entitled to receive professional advice on an ongoing basis, and in timeframes that minimize delays to the postgraduate research program.

Universities are funded to provide research supervision, and supervisors need to ensure that they are performing that task as intended – rather than using their positions of authority over a student as a means of reprioritizing their work towards other activities that may have short-term career benefits.

Allocating time to research students is, however, insufficient of itself to provide a good feedback mechanism. Key to this is understanding that criticizing a research student doesn't make them smarter or more capable – it just demoralizes them, makes them insecure in their own thought processes and less open with the supervisor.

The feedback which supervisors need to provide therefore has to be in line with the capabilities of the research student to put it to use. For example, telling a research student that a series of experiments has been poorly structured and ill-conceived may be of little value if a research student does not grasp the concepts of experimental design as they pertain to his/her particular field of research. Telling a research student that an analysis of results is flawed is unhelpful unless the student has a sound grasp of statistics. To this end, a supervisor needs to:

- Maintain ongoing vigilance of the research student's activities.
- Endeavor to assess where the research student's strengths and limitations lie.
- Look towards providing support on a pro-active basis for areas that appear to be weak – for example, having the student sit in on a relevant statistics course if that is determined to be an area of weakness.

More than this, it is important that the supervisor provides useful (i.e., constructive) feedback when the need arises during meetings with the student. It can be tempting for novice supervisors to provide feedback in the context of pointing out what the student has done incorrectly. Rarely is this sort of feedback useful in its own right, and if its only purpose is to provide the supervisor with an air of superiority, then it is completely counterproductive.

A research supervisor should have a good grasp of whether a research student has performed an allocated task well or not. This is something which the supervisor needs to process internally, rather than to vocalize to the student. The end objective is not to tell a student what he/she has done incorrectly but, having assessed the problem, to explain to the student what needs to be done in order to get things back on track.

When a research student has performed a task poorly or incorrectly, then time and resources have been lost and it is important, in the context of feedback, to enunciate the consequences of those losses and what can be done to remediate them. For example:

> *"...This is where we are now, and here is where we should be. In order to get back on track, you will need to provide me with a plan on how you intend to get these new experiments completed by this date. The cost of the ones you have already performed will need to come out of the budget, and so you will need to provide a revised budget in order to bring the costings back into line."*

It is important not to dwell on errors but to focus on the remediation strategy. The old adage that nobody wants to listen to problems is particularly relevant here – the focus needs to be upon solutions, not regurgitating history or looking at punitive measures. Nevertheless, universities and their research groups are time and resource constrained environments, so any remediation strategy should self-evidently reveal to the student the latent penalties or sanctions that are incurred as a result of any errors/shortcomings that have arisen.

In exposing the time and resource constraints that rein in the project as a result of the student's performance, it is also important not to inadvertently leave the student with the impression that one is incentivizing a *corner-cutting* approach to research – in other words, letting the student think that second-rate research methods are a sensible solution to the problem.

In the matter of avoiding blame allocation, it may also be useful, particularly when things have gone awry, for the supervisor to use the collective pronoun, *we*, rather than *you*. This avoids victimizing the student, and demonstrates that the supervisor and student are working with one another, rather than against one another.

Apart from demoralizing a student, negative (i.e., unconstructive) feedback is also an unproductive pathway from a management perspective. If the student feels that he/she is going to be victimized or humiliated by the supervisor each time he/she makes a mistake or error of judgment, then the student is going to become reluctant to bring any dubious issues to the supervisor for discussion – in the end, the dubious issues may compound into serious and/or irreversible problems that may doom the research program to failure.

A good supervisor should welcome any attempts by a research student to bring problems to the table so that they can be discussed without blame or rancor, and with a view to genuinely achieving a better outcome for the research program.

8
Planning the Postgraduate Research Program

8.1 Introduction

President Dwight Eisenhower often recited a statement which he had heard in the army in relation to planning – specifically that *(Eisenhower.archives.gov, 2015)*:

> *"Plans are worthless. Planning is everything."*

The army dictum is as relevant to modern postgraduate research as it was to the army decades (or perhaps centuries) ago. There is so much knowledge to be gained simply by going through the planning process, that what documents eventuate from it are neither here nor there.

There is one proviso to the dictum however – and that is that the planning process must be a genuine process – and not just a *ticket-punching* exercise. When planners put forward timeframes for elements within the plan, those timeframes need to be the best possible estimate of what is required – not unrealistically short, merely to impress – and not unrealistically long, in order to provide padding in the process.

Many academic purists become aghast at the idea of formally planning a postgraduate research program, and work on the assumption that *it will take as long as it takes* in order to make a significant contribution to knowledge. Laudable though this commitment may be from a purist perspective, the reality is that most postgraduate research programs are time-limited. Moreover, as an apprenticeship (prelude) for research or professional life elsewhere, students naturally want to complete their studies as efficiently as possible in order to move on and make contributions (and an income) elsewhere.

There is also a case to be made that research students, like gases, will fill available space with activity and so, if they are allocated a month to complete a research task, it will take a month – if they are allocated two months, it will

take two months. There need to be downward pressures applied to ensure that something which can be achieved in a month is achieved in a month. There are two kinds of pressures:
- Internal – self-motivation on the part of the student.
- External – formal time boundaries imposed by the supervisor.

Many research students will not previously have worked in a high-pressure professional environment, where a deadline is a deadline – and there is no possibility of extension. Many students may not have had to dramatically increase their workload, and cut out recreational/personal activities in order to achieve a result. At an undergraduate level, and particularly because postgraduate students come from the top end of the cohort, it may be that research students have been able to breeze through their studies with minimal effort or sacrifice.

Sometimes, research students consider recreational and personal activities (*me time*) as an integral part of their natural work activity. Part of the lesson that has to be learned during postgraduate research is that in the professional world, none of these models may be valid. Importantly, this learning about deadlines needs to come about through project planning.

It can be particularly useful to get the postgraduate research student to develop his/her own research plan as part of an iterative/consultative process with the supervisor. This activity is best initiated during early meetings with the research student.

8.2 Understanding a Deadline

Most adults, and almost all professionals, would claim to understand the meaning of a deadline. The problem is that in an academic environment, where research problems and approaches can be abstract and nebulous, staff and students often don't genuinely believe in them. Planning has very little value when those formulating the plans – particularly the research supervisors – do so with a lackadaisical mindset:

> *"We'll just say one month for these experiments, and if they take longer, well they just take longer..."*

This is where postgraduate research planning goes awry, or degenerates into little more than a worthless bureaucratic activity – producing a document in which neither party believes, and with which neither party intends to comply. In other words, a complete time-wasting exercise.

8.2 Understanding a Deadline

The mindset for genuine project planning has to be completely different:

"We'll say one month for these experiments. I'll check your progress after two weeks, and if the results are falling behind then you will need to work evenings and weekends. If we are still behind after the third week, then we'll have to look at hiring additional technical staff to support the work."

The difference between the former and the latter approach is stark. In the former approach, there is no real commitment on the part of the supervisor to ever meet the deadline. Any intelligent research student could decode the subtext of this into:

"Take as long as you need to do the experiments and just write down one month on the plan for the sake of putting down a notional time."

In the latter approach, the deadline is set in stone, and has the imprimatur and commitment of the supervisor. Workloads and additional resources are adjusted to ensure that it is met.

A supervisor needs to determine which approach he/she and the research student intend to pursue – if it is the former, then much time and energy can be saved by not planning the project at all. If it is the latter, then the supervisor needs to:

(i) Ensure that timelines in the plan are feasible.
(ii) Monitor/oversight the student's execution of the plan.
(iii) Identify when/if a student is falling behind.
(iv) Consult with the student to ensure that he/she is actively increasing workload if falling behind.
(v) Be prepared to inject additional resources into the project if it becomes apparent that a student is working to maximum capacity and still falling behind schedule.

The key point here is that planning is about more than just creating a document – which superficially looks convincing – and then retrospectively accepting that it was too difficult and ambitious to achieve.
Planning requires:

- Genuine commitment and honesty on the part of the supervisor and the research student.
- Real oversight (not just superficial glimpses) on the part of the supervisor.

148 *Planning the Postgraduate Research Program*

- Proactive correction of any lapsing elements of the schedule through additional workload or resources.

Academics in general don't like being constrained. Often, they will cite the complexities of their research work as reasons why rigid planning rules won't work – *too many unknowns*. However, this is a specious argument to the extent that any major project – be it,

- The development and manufacture of a new vehicle model.
- Construction of a major highway or shopping mall.
- Development of a new aircraft,

has far more unknown elements than a university research project. And yet, these elements can regularly be managed and projects and outcomes delivered *on time*. In a commercial sense, a failure to deliver on deadline can be a catastrophic professional outcome – and a costly one. The majority of postgraduate research students will ultimately end up in some form of commercial or government career, and it is therefore best to develop planning rigor as early as possible.

8.3 Basic Elements of a Postgraduate Research Program

The differences between postgraduate research programs, across disciplines, fields and institutions, can be significant. However, there are numerous, common elements to all postgraduate research. Some of the basic ones are shown in Table 8.1.

In the context of postgraduate research planning activities, it may be particularly helpful for a supervisor to compile a more detailed list of activities that relate specifically to his/her research field. When conducting initial meetings with the research student, the supervisor can call upon the compiled list in order to assist the student with his/her planning.

Invariably, some of the elements in the research program will have ill-defined start and end times but this should not diminish any attempt to ensure that reasonable efforts are made to estimate appropriate event durations and work towards achieving them.

In many institutions, the final assessment of the postgraduate research program is undertaken by peers, external to the university in which the research is conducted. The examination process then becomes one of the most ill-defined time durations in the overall planning process – albeit one which is outside the scope of the supervisor and student's capacity to change.

8.3 Basic Elements of a Postgraduate Research Program

Table 8.1 Basic elements and time considerations is postgraduate research

Element	Objective	Time Considerations
Literature Review	To determine the state of knowledge in the field as presented by noted scholars/peers	Maximum effort at beginning of program, with ongoing background activity throughout the entire postgraduate project
Formal Determination/Establishment of Hypothesis	Based on a reading of scholarly literature/opinions, to identify knowledge gaps that can be tested as an open hypothesis which does not predispose the research to a particular outcome	First priority to be resolved as quickly as possible after extensive literature review
Preparation of Research Methodology	Development of a methodology in line with scholarly/peer approaches in the field	Active components of research program cannot commence until methodology is established
Design of experiments or instruments to test hypothesis	Development of systematic tools to determine whether the "null hypothesis" can be disproven	Formal experimentation is contingent on the completion of this design
Experimentation or conduct of investigation	To undertake the actual work required to test the stated hypothesis	Generally constitutes the bulk of the postgraduate research program in time
Systematic aggregation of results or information	To bring together the results gathered from experiments or other instruments in a concise, systematic format	Results are better aggregated and analyzed on a regular, ongoing basis or after completion of a milestone
Detailed analysis of results	To assess how detailed results either support or contradict the stated hypothesis	
Publication of results/findings in scholarly Literature	To subject research results to an external peer review as a first "impartial" test of the validity of the approach and outcomes	Preparation of research paper can take weeks of effort. Journal peer review process can take months
Publication of complete body of work in Dissertation	To completely document the research investigation from initial literature review through to self-assessment of strengths, limitations and scope for future research	Research dissertation should be under development from the beginning of the research program and ongoing throughout
Examination Process	To subject to the entire research investigation to external, independent peer review	Maximum length of examination process needs to be determined and factored into plans.

A supervisor (or even the university itself) has little practical influence over how long the examination process may take. Some institutions insist that external examiners complete their assessments within a limited timeframe but, realistically, they have little influence over what ultimately happens. The problem is that it is difficult to attract suitably qualified examiners, and each generally needs to be scrutinized by a university research committee before appointment. If an examiner is unable to fulfill his/her duties on time, it is often impractical to go through the process of selecting an alternative examiner and then repeating the process. The end result is that there may be considerable variability in the duration of the examination – and this may have a profound impact upon a research candidate's ability to move on to a new position.

In the context of planning, given an inability to control the examination process, it may be more practical to omit this from the research student's overall plan, notwithstanding the fact that the student needs to be aware of the idiosyncratic nature of the process.

8.4 The Planning Process

There are various project management software packages available to assist in the development of a postgraduate research plan. The reality, however, is that in the case of an individual research student, the critical path in the process is predominantly dependent upon that research student – as are the bulk of the activities therein. For these reasons, students can expend (waste) considerable time learning to use (yet another) software package which adds little value to the postgraduate planning process. Unless a supervisor believes that there are extra-curricula benefits to using a specialized package, then encouraging the use of a generic spreadsheet package to achieve a more efficacious outcome may be beneficial – and avoid time-wasting on technologies which are over-kill for a very basic postgraduate project plan.

It is often the case that research students seriously underestimate the amount of time that is actually available to undertake a postgraduate research program. In particular, the notion of having several years to complete an investigation is – from the perspective of a research student (who is often only 22 years of age) – an inordinately large amount of time. After all, it is around a seventh of a student's entire lifespan to that point.

The key point for the supervisor to get across to the research student during early meetings – and the planning process – is that, firstly, even though the research program may span several years, there is no real slack built into the process. Secondly, in order to get an insight into the tight time-frames associated with the program, the research student needs to learn to *backward-schedule* all his/her activities from the expected completion date of the project. For example, if the candidature duration is three years (to submission of final thesis), then:

- Week 156 of the candidature is the reference point from which all work is backward scheduled.
- This may mean, for example, that the supervisor needs to complete his/her final review of the thesis by Week 152.
- This may, in turn, mean that the student's complete draft needs to be completed by Week 146.
- This may require all results to be analyzed by Week 138.
- Actual experimentation/investigation may need to be completed by Week 130,

and so on.

When a research project of several years is divided into weeks, and allowances are made for a supervisor's time to read and assess submissions, students can be genuinely taken aback – or even alarmed – by how little time is actually available to them to complete their investigation.

The other point that needs to be made to the student is that the tasks associated with postgraduate research are not always contiguous – many of the tasks overlap and some need to be performed simultaneously in order to enable the project to finish on time.

Even in the absence of a formal project management software tool, the basic mapping process is very straightforward. The horizontal axis of the management plan represents time and the vertical axis relates to specific activities that need to be performed. For postgraduate research programs, there is little value in resolving time into units of less than a week – so a three year program would have 156 time units in which activities need to be scheduled.

Figure 8.1 shows a very basic (sample) project management chart with broad activities listed over a 156 week period. In this simple plan the details of the investigative process and thesis development have been left open-ended because of the broad spectrum of possible activities that might take place in any given research program. A research student would need to develop his/her own chart with a higher level of resolution for these activities and milestones as they relate to issues, such as:

- Thesis development (e.g., which chapters will be completed and when).
- Design/development of specific apparatus/equipment.
- Execution of the specific methodology (e.g., which experiments will be completed and when).

152 *Planning the Postgraduate Research Program*

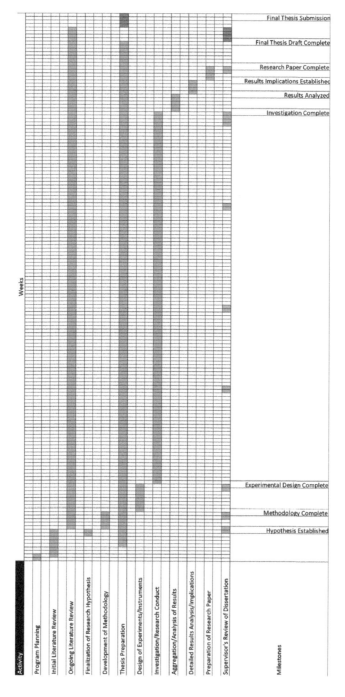

Figure 8.1 Basic elements of project management chart.

Again, however, it needs to be stated that there is little value in expending time on this activity unless both the supervisor and the research student view the milestone dates and activity durations as hard and fast. It is difficult enough for research students to meet milestone targets when they are viewed psychologically as immovable events. Where they are viewed as little more than moveable *guestimates*, the planning process loses much, if not all, of its value.

Milestone dates should only be moved as an option of absolute last resort – when all other options of increasing resources to meet them have been expended – and not as a first option, or path of least resistance, just because meeting them is difficult.

8.5 Using the Planning Process

Creating plans is one thing – actually using them is another matter entirely. If the project plan is to have genuine value for both the supervisor and the research student, then it should form the basis for discussions at all formal meetings. Specifically, a supervisor needs the student to explain:

- How he/she is performing relative to the plan.
- If the student is falling behind and may not meet a deadline, what additional resources need to be deployed (including having the student working longer hours)?
- Do changes need to be made to the plan – *vis a vis* new activities or deletion of existing activities?
- Do plan changes also necessitate resourcing changes for the project (e.g., will development of additional apparatus require additional technical support?)?

An obvious question that also needs to be asked is what needs to be done in the event that the plan is completely flawed from the beginning? The answer is that there really isn't sufficient scope in a postgraduate program to have such major structural flaws in the plan in the first instance. Planning a postgraduate program isn't like planning the development of a new passenger aircraft – involving thousands of staff, numerous partners and hundreds of variables. A postgraduate research plan has:

- An overall duration generally defined by the university.
- A single entity (student) responsible for the bulk of all activities.

- A single entity (supervisor) responsible for overall management.
- Well-defined outcomes (e.g., thesis, research papers).

Of course there are always uncertainties with knowledge creation, and even accessing resources and so on, but these are not major structural issues within the plan. For these reasons, the planning process should be straightforward and so too should be its execution.

9

Conflict Resolution in Postgraduate Research

9.1 Introduction

In 1765, Samuel Johnson wrote the following words about the sorts of disputes that arise between academic commentators on classical literature (scholiasts) *(Johnson, 1765)*:

> *"It is not easy to discover from what cause the acrimony of a scholiast can naturally proceed. The subjects to be discussed by him are of very small importance; they involve neither property nor liberty; nor favour the interest of sect or party. The various readings of copies, and different interpretations of a passage, seem to be questions that might exercise the wit, without engaging the passions.*
>
> *But whether it be, that small things make mean men proud, and vanity catches small occasions; or that all contrariety of opinion, even in those that can defend it no longer, makes proud men angry; there is often found in commentaries a spontaneous strain of invective and contempt, more eager and venomous than is vented by the most furious controvertist in politicks against those whom he is hired to defame."*

Subsequent variations to this theme – particularly those by Henry Kissinger – have led to the modern day observation that disputes within academia tend to be profoundly bitter, particularly given that the issues surrounding them are often of little consequence to those outside the warring parties.

This is not to say that all conflicts in universities are frivolous, and there is always a legitimate case for conflict when the issues at stake include:

156 *Conflict Resolution in Postgraduate Research*

- Academic misfeasance.
- Perversion of knowledge.
- Misuse of institutional funds or resources.
- Mistreatment of (discrimination towards) staff or students – including harassment and bullying.
- Abuse or mistreatment of laboratory animals.
- Endangerment of health or safety.

Clearly, these are not issues that can be negotiated into resolution, and some formal action is required to remedy them. However, these sorts of issues tend to be in the minority – the majority of university disputes fall into the category of *large disputes over small issues*. For these matters, it is important that research supervisors learn to recognize when a dispute is about to occur (or is occurring), and how to act to ensure that it does not escalate unnecessarily.

In Section 4.5, the issue of conflict resolution, as it pertains to the relationship between the research student and supervisor, was discussed. Herein, the factors involved in resolving conflicts with other parties, including:

- University committees.
- Departmental heads, faculty deans, institutional heads.
- Technical staff.
- External research partners,

are examined.

The key to resolving conflict issues in universities is an understanding that:

- Disputes and arguments arise from time to time and are a normal part of the professional environment.
- At some point, disputes come to an end and, generally, a normal working relationship needs to be resumed between the parties.
- In managing a dispute, it is important to recognize and plan for what will happen at the end of the dispute and how goodwill will be resumed.
- If the resolution of a dispute involves the creation of a winner and a loser then, in reality, both parties are losers because it is difficult to restore goodwill in such an environment.
- Professionals need to be capable of completely separating workplace disputes from personal feelings and egos.
- Most disputes in universities need to be resolved with some degree of compromise from both parties.

Table 9.1 shows the various stages of conflicts in the context of universities.

Table 9.1 Stages of conflict in disputes

Stage	Phase	Issues
1	Initial Dispute	An initial dispute may be a minor flare-up over a small matter – for example, office space, IT resources, etc.
2	Escalation	Parties to the dispute take an entrenched position in the conflict and it becomes more serious
3	Injection of Personal Issues	The initial dispute often gets lost in personal antipathies that are exposed as the dispute escalates
4	Loss of Productivity	Parties to the dispute end up wasting time and resources on correspondence to each other (often achieving little more than restating an entrenched viewpoint) and in endeavoring to garner support for their cause at a higher level – the cost of productivity loss (including wasted salaries) can be significantly larger than the issues originally at stake in the dispute
5	Mediation	As the dispute escalates, it is often necessary for a more senior officer of the university to act to mitigate the dispute or to simply rule on a resolution
6	Compromise, Win or Lose	The dispute is settled – either through both sides compromising or one side winning and the other losing
7	Restoration of Relationships	Relationships between the disputing parties need to be restored to stop departmental dysfunction

A good way to view any conflict with another party in the university system is to draw up a mental flow-chart which maps out all the possible responses from the other party at each stage in the conflict. Recognizing that the last step in the flow-chart is the restoration of the working relationship helps to make one significantly more cautious in tackling the dispute at every stage.

There are three common mistakes that arise with professional novices in disputes:

(i) Making them a personal issue between individuals. This is a luxury that is seldom accorded to senior professionals, and it is a lesson that novice professionals need to learn very quickly.
(ii) Individuals in a dispute decide that they are *right* and the other party is *wrong*. In the workplace context, it is generally irrelevant who is *right* and who is *wrong*, because inevitably both parties to a dispute believe that they are *right* – otherwise there would be no dispute.
(iii) Pursuing the naive idea that one can convince another party that they are *wrong* or, even more naive, that one can somehow change the persona of the other party such that they fall into line with one's own views.

There is an old saying that *nobody ever changed their mind as a result of losing an argument*. True or not, this is a useful basis upon which to look at dispute resolution. As a professional, the objective of dispute resolution is not so grand as to win an argument, or to convince another to change his/her mind, or even change what the other party believes or how he/she operates. The objective is to find a professional solution that both parties in the dispute can live with – and then move forward.

Nevertheless, it is usually the case that when disputes degenerate into questions of *right* or *wrong*, a professional dispute can become personal – and when it does, it is all the more difficult to resolve.

It is also critically important to understand – before a dispute gets personal – that it is exceedingly rare, in any organization (especially a university), for an individual to have sufficient authority to resolve disputes by fiat. One of the rare professional exceptions to this would be in a company where an individual owner has complete authority. In a university, however, regardless of who attempts to issue a decree, the other party generally has recourse to redress from a higher authority – which may undermine that decree and severely undermine the credibility of the person making it.

The obvious solution then is *not* to issue unilateral decrees to resolve disputes.

Given all the things that should not be done when disputes arise, the obvious question is what should be done? The answer, which should be self-evident, but rarely followed once a dispute has degenerated into the personal, is to find out what the other party wants done in order to resolve the problem. Unfortunately, in many disputes, the parties have a habit of restating their problems more loudly, rather than stating their preferred solution. Desmond Tutu, in a speech to the Nelson Mandela Foundation in 2004, noted,

"Don't raise your voice, improve your argument."

To this end, the first steps towards resolving conflicts should involve:

- A statement of each party's perspective on the problem, rather than an iterative restatement of the problem itself.
- A statement of the external constraints under which each party operates (e.g., time, resources, funding, etc.) and which pertain to the problem.
- A request for the opposing party to put forward their version of an ideal resolution to the problem in light of the constraints that have been enunciated.

It is often the case that what initially appear to be unreasonable demands from another party in a conflict have only been made as an ambit claim. When the other party is asked to put forward a resolution, in light of constraints, rather than as an open-ended restatement of the problem, quite possibly the initial demands will be moderated and provide scope for compromise. For example, consider the following approach:

> *"I've told you all the constraints that we have to work under, and I don't have the authority or resources to change those. Now, given those constraints, tell me what you think would be the best possible solution that we can work towards from your perspective."*

In other words, if one individual believes that another has started a dispute, then it is important to encourage the person who started it to put forward a mechanism for its resolution. In so doing, one has to be prepared to avoid intransigence – there is no point asking another party for a solution if there is no intention of bending to genuinely consider the proposal as a possible solution.

Importantly, once derogatory remarks are enunciated or, worse still, put in writing, it is very difficult to withdraw them or have the other party put them to one side in order to restore a working relationship once the dispute is resolved.

The key lesson here is that, at all times, it is important to remember that work is work and personal is personal – and the two should never be confused. A dispute in the workplace is business – and it needs to be resolved in a professional, business-like manner and not through personal antipathies.

9.2 Jurisdiction and Scope of Conflicts

It is important that time is not wasted on disputes that are outside the control of both of the parties involved in them. For example, if there are space constraints inside a physical building, and the only solution is to erect a new building, then it is pointless to have junior staff members bickering over the need for that new building – when neither party has authority or resources to do anything about it. Similarly, if one party alleges that another has acted unethically, then it is pointless having a personal argument between the two if such matters are normally heard, as a matter of course, by a special university research ethics committee.

In some cases, the simplest dispute resolution mechanism is therefore to pass the issue off (up) to those who have the authority to deal with it.

For some conflicts, there will also be a need to defer to more senior staff or to specialists in order to respond to a party making a dispute. For example, an external research partner may start a dispute over the distribution of intellectual property (IP). This may require deferment to the university's legal department or research commercialization department. Moreover, in trying to second-guess or preempt how these departments may respond, an individual may inadvertently escalate the dispute unnecessarily.

For research supervisors, even a dispute with a research student may require deferment to other parties if it is sufficiently serious. For example, if a research student accuses the supervisor of acting unethically – or, if the supervisor accuses a research student of acting unethically. Once preliminary discussions between the supervisor and student have been completed, if there is no goodwill, then such an issue needs to be escalated to a higher authority. And, it is important for the supervisor not to commit himself/herself to a particular course of action if the matter is being escalated.

9.3 Not All Conflicts Can or Should Be Resolved

9.3.1 Basic Issues

An important question to consider is whether all conflicts in the university environment can be resolved through compromise and goodwill. The answer to this is an emphatic no – and further, some conflicts need to be escalated in a formal sense as a matter of professional conduct. Specifically, the following issues are instances where resolution by negotiation is generally not an option:

- Academic misfeasance.
- Perversion of knowledge.
- Misuse of institutional funds or resources.
- Mistreatment of (discrimination towards) staff, students – including harassment and bullying.
- Mistreatment of laboratory animals.
- Endangerment of health or safety.

These issues are not commonplace in the university environment but neither are they exceedingly rare, and most academics will need to deal with them at some point in their careers. Specifically, academics need to consider that:

(i) In some jurisdictions, these issues have serious legal (and often criminal) ramifications that can lead to sanctions including custodial prison sentences.

9.3 Not All Conflicts Can or Should Be Resolved

(ii) Eventually these issues will become public – or get raised at a higher level – so attempting to *sweep them under the carpet* is not an option – they need to be dealt with as soon as they come to light.

(iii) When such issues do become public – or get raised at a higher level – those who were aware of them but did not act to prevent them may be deemed as culpable – morally, legally and professionally – as those who were the protagonists in them.

These disputes differ from the usual day to day conflicts that emerge in academia in many ways, not the least of which is because it is highly unlikely that a normal working relationship can be resumed with the opposing party once the dispute is settled.

Needless to say then, these are particularly difficult and traumatic issues to manage – particularly if a junior staff member becomes aware of a more senior staff member having some involvement in wrongdoing. Further, there may come a time where a staff member – including the one reporting the wrongdoings – may have to resign his/her position – whether this seems fair or not. Such are the high stakes in these serious issues.

Of particular significance to this discussion is the famed quote from John Stuart Mill *(Robson, 1984)* in his address to the University of St. Andrews in 1867,

> *"Let not any one pacify his conscience by the delusion that he can do no harm if he takes no part, and forms no opinion. Bad men need nothing more to compass their ends, than that good men should look on and do nothing."*

Culpability for wrongdoing in the university sector can also be enacted in national or regional law as much as such colorful prose – particularly because universities are often dealing with government funds which are often protected in the scope of their usage by legislation. Academics should therefore consider legal and professional ramifications very carefully before deciding upon a specific course of action.

Finally, all academics need to be aware that failing to confront issues of wrongdoing is not a solution – and nor will it make the problem go away. Perpetrators are often emboldened by a lack of action on the part of other colleagues and, far from desisting in these activities, they are likely to escalate them.

In this section, the emphasis is in dealing with the issue of academic misfeasance in terms of specific dispute resolution. Chapter 13 of this book will deal with the issue of academic misfeasance in a broader sense.

9.3.2 Managing Disputes Relating to Misfeasance/Wrongdoing from a Senior Staff Member

The starting point for dealing with issues relating to misfeasance/wrongdoing in the university sector is that people need to be aware of:

- University regulations/procedures.
- National/regional legislation pertaining to issues that impinge upon academic activity.
- Health and safety requirements in the institution.
- Basic issues of ethics as they pertain to research process, publication, conflict of interest, etc.
- Basic issues/rules/procedures/legislation pertaining to discrimination, harassment, bullying, etc.
- Basic issues/procedures relating to financial practice as it pertains to:
    - The expenditure of university funds.
    - Conflicts of interest in expenditure of university funds.

It may seem unfair that someone taking on the role of a research supervisor – perhaps even as a novice academic – should need to have an appreciation and awareness of such broad-ranging issues, but such is the nature of working in a complex, modern work environment. The harsh reality is that a lack of understanding of these issues is not deemed to be a reasonable excuse in the event that an episode of misfeasance/wrongdoing becomes exposed.

Those undertaking research supervision and managing university research funds should also be aware that modern accounting software systems and auditing processes within institutions are designed to pick up irregularities in expenditure and invoicing. Anomalous items can be flagged and formally investigated through audit. In particular, the following items are commonly scanned:

- Inappropriate expenditure of funding against unexpected cost codes (e.g., use of research grant funding for personal entertainment or travel purposes).
- Inappropriate invoicing of the institution (e.g., establishment of artificial companies by staff members for the purposes of invoicing the university).
- Inappropriate authorizations of expenditure with conflict of interest (e.g., a staff member authorizing payment of an invoice from a company in which he/she (or a relative or spouse) is an owner or shareholder.

With these points in mind, the most obvious method for managing disputes relating to misfeasance/wrongdoing is for a person never to personally engage

9.3 Not All Conflicts Can or Should Be Resolved

in any such activities in the first instance. Nevertheless, on occasion, a staff member may be exposed to the sorts of activities enunciated above as a result of the actions of another staff member. Worse still, it may be the case that a senior staff member asks a junior staff member to engage in some activity that the junior staff member knows is unlawful or unethical. This places the junior staff member into a diabolical predicament. How then can the junior staff member manage such a conflict?

The following approach may be useful:

(i) Never rely on the spoken word when asked by a more senior staff member to carry out an activity which may be considered to be unethical or unlawful – insist that all discussions with the senior staff member are put in writing.
(ii) If the senior staff member refuses to put instructions in writing but insists on verbal processes (as is likely to be the case if they are aware of their own misfeasance), then the junior staff member needs to echo any verbal requests – in writing – back to the senior staff member, and ask the senior staff member to verify – in writing – that these are his/her instructions.
(iii) If the senior staff member refuses to confirm – in writing – his/her requests, the junior staff member should seek formal advice at a more senior university level before acceding to any request.
(iv) An academic's first responsibilities are to federal/regional laws and to the institution itself – not to any colleagues within it – regardless of their seniority.
(v) Retain off-site copies of all hard and electronic correspondence with the senior staff member – and all other documentary trails, particularly financials.
(vi) When there is sufficient documentary evidence of wrongdoing, formally approach either a senior officeholder of the university or their nominated representative (e.g., head of audit; head of ethics committee, etc.).
(vii) Complaints to university officeholders should be entirely based upon facts – preferably documentary facts (evidence) – rather than personal recollections of conversations.
(viii) Do not confuse personal opinions or suspicions with facts – serious claims against other staff members need to be based on hard, irrefutable evidence and not just accusations or hearsay.
(ix) Always be aware that the laws of slander and libel may apply to any individual who makes unfounded (and, in some jurisdictions, even founded) accusations against the character of another person – this is all

the more significant if accusations are made against the character of an otherwise eminent person.
(x) Do not discuss or circulate correspondence or other evidence of potential wrongdoing with colleagues – this may be deemed to be slander or libel. Any disclosures should be treated as privileged, confidential information only discussed between the complainant and formally appointed officeholders/investigators of the university.
(xi) If senior officeholders of the university are not prepared to deal with misfeasance/wrongdoing (e.g., for fear of damage to the institutional reputation, or because the staff member involved is an eminent person), the junior staff member needs to go outside the university to other suitable bodies (e.g., police, federal/regional government legislative representative, national/regional ombudsman, etc.).

Needless to say, the stakes in such disputes are extremely high, and raising wrongdoing, misfeasance or broader corruption within an institution could have serious professional consequences for the person raising them – particularly because of retribution from those involved or their colleagues.

A novice staff member should also consider that the trail of wrongdoing may not stop with the staff member with which he/she has had an altercation – it may extend upwards into more senior management. It may also be the case that senior management is already aware of the ongoing misdemeanors and has been deliberately avoiding confronting them for fear of the public consequences. If this is the case, then the novice staff member needs to consider his/her position within the institution.

9.3.3 Resignation from the University

Universities are the gatekeepers of knowledge, and all staff have an implicit (and generally explicit) onus upon them to ensure that activities within the institution are undertaken ethically, professionally, and in a non-discriminatory manner in a spirit of fairness. This gatekeeping role is perhaps the single most important activity that any academic will ever undertake.

It is critical, therefore, that academics do not engage in the self-delusion that their own research is so critical to the world that the gatekeeping role can be relegated to secondary importance – and that they can therefore *look the other way* when they see wrongdoing, for fear that exposing it will damage their own career. The reality is that, sooner or later, the process of knowledge discovery will inevitably take its natural course, regardless of whether any individual academic (even at the highest level) participates – so all academic careers are ultimately expendable. However, the buck for the gatekeeping role stops abruptly with each and every individual. It is also the case that – eventually – the truth will out and, if there is wrongdoing, an individual will either be branded as part of the problem or part of the solution.

Generally, the final and most serious act in the gatekeeping role of an academic is to formally resign from the institution when it becomes apparent that misfeasance is an ongoing problem, and all attempts to prevent it from happening within the institution have failed.

The reality is that universities are large organizations and, as such, suffer from the common trait wherein senior management can fail to confront wrongdoing/corruption until it reaches crisis point, and action becomes unavoidable. All too often, a failure to confront wrongdoing and corruption in academia is rationalized by the notion of *protecting the institution's reputation*. The reality is that such rationalization is more likely a result of staff avoiding the ugliness of confrontation as well as sparing their own personal reputations and careers.

If it is the case that misfeasance in an institution is widespread, and senior management do not wish it to be exposed, then an unfortunate common practice is to *close ranks* at the highest level and declare that instances of misfeasance have been reviewed and all allegations have been settled. In such a scenario, where a staff member is not satisfied that this is genuinely the case, then there are several options:

- Seek independent legal counsel on the matter to ensure that continuing to work in such an environment will not lead to potential legal or professional sanctions.
- Take the matter outside the university management's control – to an appropriate government legislative representative or ombudsman.
- Take the matter to an investigative journalist for further analysis – however, this creates the risk of exposing the staff member to defamation and/or action from the university for inappropriate disclosure of confidential in-house information.
- Resign from the institution, formally documenting all evidence that has come to light and presenting it formally – in writing – to appropriate senior officeholders of the institution and, if necessary, government regulatory officials.

It also needs to be remembered that the notion of allowing wrongdoing to continue because,

> *"I'm not the one involved in the wrongdoing – why should I be the one to resign when the problem is in more senior ranks?"*

is not a professional solution. Ultimately, when the wrongdoing is exposed – and it will be – the bystander staff member may also be forced to resign anyway – and, depending upon the nature of the offences committed by others – may also face serious legal sanctions.

Resignation is naturally the absolute last resort in the process of confronting wrongdoing within an institution, and it will have a profound impact upon the career of the person resigning. It therefore requires considerable thought and reflection. However, for an individual with professional integrity and ethics, the option of staying on in an environment rife with wrongdoing may be a far more stressful option in the long term.

It is difficult to think clearly and rationally in the midst of a professional crisis, where:

- There may be severe pressure exerted from the upper echelons of the university, insistent that everything has been checked and resolved.
- A decision to resign may completely negate an entire lifetime's worth of aspirations.

Every academic, at the start of his/her career, may therefore benefit from considering the following points:

- At some point in the career, a professional crisis relating to wrongdoing in the work environment is likely to erupt.
- A decision will need to be made in relation to confronting the crisis head-on, or living with wrongdoing, in the knowledge that – left unchecked – it will not only continue but possibly escalate.
- Tackling the crisis may require a resignation and, in the worst case scenario, a possible change of career.
- Does the academic value personal integrity above career? If not, is he/she an appropriate person to be a gatekeeper of knowledge?

9.3.4 Managing Disputes Relating to Misfeasance/Wrongdoing from a Research Student

Managing conflicts that arise when a supervisor becomes aware of misfeasance/wrongdoing on the part of his/her research student is significantly different to managing such occurrences from senior staff. To begin with, because research students have limited authority within the university, their capacity for wrongdoing is also limited, relative to senior officeholders of the institution. Nevertheless, there are occasions where a research student may be suspected of some misfeasance and it is left to the supervisor to initiate a process of redress.

9.3 Not All Conflicts Can or Should Be Resolved

The process commences by:

- Ascertaining the facts (documents and hard information) that are relevant (as opposed to personal opinions, hearsay or accusations).
- Determining whether the facts actually point towards misfeasance/wrongdoing.
- Making a judgment as to whether the misfeasance/wrongdoing has occurred as an intentional act, or as a result of naivety or lack of professional experience (e.g., failing to appropriately cite the original author in a publication) on the part of the student.
- Determining whether the misfeasance/wrongdoing has occurred as a result of a failure on the part of the supervisor (e.g., failing to explain how certain activities should be handled) or because of inadequate resourcing/support provided by the supervisor.
- Understanding that the research student – although technically a qualified professional – is only a trainee research professional and therefore needs to be given some latitude in consideration of irregularities that may arise.

Once this initial assessment is made, a research supervisor needs to determine whether a *prima-facie* case exists for treating a student's action as misfeasance. If the answer to this question is yes, then the supervisor needs to ensure that he/she is fully apprised of institutional procedures in relation to misconduct on the part of research students. Often it is the case that institutional misconduct definitions and remedial procedures for students are far more clearly articulated than those for staff.

The combined picture – evidence, personal assessments and institutional procedures/guidelines should then form the basis for a formal meeting with the research student. In the context of ensuring that there can be a positive, ongoing relationship with the student if the matter is resolved, the supervisor should not engage in bullying or belittlement of the student but rather present the facts as a problem that he/she and the student need to address together. If the outcome of the meeting suggests that the wrongdoing needs to be tackled systematically within university procedures, then the student needs to be advised of the process and the support mechanisms that are available to ensure a fair and just methodology.

The following points are important:

- Discussions with the student, and the issues at hand, need to be treated as privileged information which is strictly confidential and only disclosed in the formal context of university procedures,

- Any allegations made by the supervisor about the student – either to colleagues or in a more public domain – may be viewed by the student as either slander or libel and could result in litigation against the supervisor,
- Under no circumstances should a supervisor bully or harass a student, or endeavor to coerce them into admitting guilt,
- The supervisor should conduct all discussions with the student on the presumption of complete innocence, and all questions should be open and without suggestion of a predisposed opinion or process outcome.

While the process here may be less traumatic for the supervisor than a confrontation with a senior staff member, it needs to be remembered that a failure in process – that is, a failure to:

- Treat information confidentiality and protect the privacy of the student.
- Provide a fair and just process where the student has the right to natural justice – that is, to challenge inconsistencies or shortcomings, and to appeal decisions.
- Protect the student's health and welfare – particularly mental health – during a stressful time.
- Restore a harmonious working relationship with the supervisor in the event that any allegations are proven to be unsubstantiated,

may result in sanctions being enacted by the university (and/or student) *against the supervisor*.

It is important that a supervisor understands that his/her word on the subject of student misfeasance is never the final word on the issue, and universities normally have processes in place for natural justice through challenge and appeal. For this reason, it is imperative that a supervisor does not overstep his/her authority by making statements which, in the context of university procedure, have no legitimacy. For example,

- *"Once this is over, you will need to find another supervisor, because I refuse to continue on as your supervisor."*
- *"I expect you to hand me your resignation from the research program and relinquish your scholarship."*
- *"I will not allow you to use the research facilities again once this investigation is completed."*

These are all examples of overstepping authority and subsuming the role of the university and its collegiate committee system – which may have ultimate jurisdiction over all such matters.

Underlying the entire process is a need to understand that even though a student's actions may have forced a supervisor's hand in terms of applying university disciplinary processes, at the end of these processes, the supervisor may still need to:

- Restore a good relationship with the student.
- Restore the student's trust.
- Extract the best possible research outcomes from the student.
- Complete the research program as originally intended.

This cannot be achieved if a supervisor in any way damages the relationship with the student during a conflict period.

9.4 Conflicts with Committees

Universities tend to have management structures where many important decision-making activities are devolved to committees – for example, committees are generally established to oversight:

- Academic coursework programs.
- Budgets and expenditure.
- Postgraduate research candidature – particularly at Doctoral level.
- Ethics.

Each university may have hundreds of degree programs and thousands of specific fields of research, so clearly it is not practical to have a single individual, with one field of expertise, attempting to oversight all of these and make sensible decisions. Committees therefore tend to be composed of staff representing major discipline/activity groups of the institution, in order to provide a coherent decision making process that still has capacity to consider the nuances of an individual field.

It is typically the case that positions on university committees are a combination of ongoing appointments and regularly elected representatives – for example, a medical faculty may nominate some of its staff to represent the faculty's interests in an ethics committee. It is also often the case that elected positions are only held for a period (e.g., several years) and the incumbents are then replaced with newly elected staff.

The committee system therefore has many advantages over running a corporate style management structure – however, it also has numerous disadvantages, specifically:

- It is difficult to get consensus on issues.
- Committees do not lend themselves to innovation and entrepreneurship.
- Committees are dominated by procedure and bureaucracy and processing tends to be slow.
- Accountability for decisions is, in practical terms, non-existent, because responsibility is *collective* – individuals within committees are not sanctioned for promulgating ideas that are ultimately agreed to by a committee.

Notwithstanding these disadvantages, there is some merit to slowing down the academic process with bureaucracy because it instills discipline and rigor into processes related to learning and research. In many instances, it is also the complexity and duration of the bureaucratic process that acts as a deterrent to wrongdoers as much as the procedures themselves.

Unfortunately, it is also the case that committees tend to be very political in nature – each member may represent a vested interest group and, in a university system where the size of the financial pie is fixed, each time one vested interest group gets a bigger slice, another group loses some of its share.

Another point to consider is the fact that if there are personal antipathies between an academic and his/her departmental/faculty representative on a committee then that representative may be able to use the authority of the committee against the academic.

All of these considerations lead one to conclude that research supervisors need to tread very warily around university committees and avoid conflict wherever possible. This is easier said than done, particularly because of the scope and power of committees within the university decision-making system. When a committee deliberation goes against a supervisor the implications can be profound – particularly if the decision causes a change or significant delay to a proposed research program because of an issue of ethics, say.

If a conflict arises between a research supervisor and a committee, then it is especially important to avoid personal antipathies from getting enmeshed in the dispute. Each deliberation of a committee may only occur on a weekly, monthly or quarterly basis, so any side issues that are raised can result in the core of the dispute being pushed over to the next committee deliberation.

The following considerations may be helpful in resolving a dispute with a university committee – generally arising as a result of a decision which is deemed unfavorable:

(i) As a starting point to dispute resolution, it is important to understand whether or not the committee had the authority to issue the decision

which is being challenged and, further, whether it has the authority to resolve the issue. This can be determined by studying the committee charter and terms of reference.
(ii) A committee decision should only be interpreted in the context of university regulations and procedures – what may otherwise appear to be a sensible course of action is generally irrelevant.
(iii) It is important to determine if the facts before the committee were adequate in order for it to reach the conclusion desired by the applicant – if not, then clearly a rejoinder needs to contain any relevant, additional information.
(iv) A committee decision should be consistent and fair – particularly in respect of previous decisions relating to similar matters. It is important to look at the history of decision-making in the area to ensure that there is consistency.
(v) If a decision has been made in a "gray area", where university regulations and procedures can be interpreted in any number of ways, it may be helpful to consult with colleagues in different universities to ascertain if similar cases have been treated differently, and if a fair-minded, impartial observer would think the same process should be applied to the decision in question.
(vi) If there are no local precedents for comparable decisions, then it may be worthwhile to check for common, international practices in the area.

In light of these considerations, a research supervisor may need to make a formal rejoinder to the committee, which courteously outlines any irregularities or inconsistencies that may lead to a different consideration of the issue.

It is also not unreasonable for a supervisor to approach the chair of the relevant committee to discuss the decision and determine what course of action (i.e., procedure) needs to be pursued in order to achieve a different outcome.

9.5 Conflicts Involving Departmental/Faculty Resources and Technical Support

In a large, complex organization, such as a university, where annual budgets are often measured in billions of dollars, it is not uncommon for students and inexperienced staff to believe that there is a surfeit of riches that can rain down from above in order to support any research program or learning activity. The reality is that no matter how large the institutional budget, there is never sufficient funding to enable every student and staff member to fund every item necessary for learning and research. Compromises inevitably have to be made.

Typical conflict points that arise in the context of postgraduate research supervision include:

- Office and laboratory space for staff and students.
- Funding for conferences.
- Funding for research incidentals (e.g., purchase of specific journals).
- Information Technology (IT) support.
- Discrepancies in the treatment of various staff (e.g., preferential treatment of new or eminent academics).
- Workload allocations and balance between teaching and research activities.

All of these issues have the potential to (and often do) flare up into large disputes which – in the final analysis – may be outside the control of the staff member and his/her line manager (e.g., head of department or faculty dean).

As with all disputes, the starting assumption needs to be that, at some point, the dispute will end and a good working relationship will need to be resumed. It is therefore imperative that nothing is done during the dispute to damage any future working relationship. It is also the case that these sorts of disputes are with more senior staff who are in a management role and will inevitably have some impact on future career aspirations – so, respect and caution are paramount.

It is particularly important here to recognize that, unlike disputes with committees, these sorts of disputes are generally between individuals, so the starting point for their resolution is to meet with the other party and have a professional dialog. Academics should avoid putting such a dispute in writing in the first instance. Generally, this only inflames a situation and forces the other party to formalize discussions in the context of a university procedural framework, rather than in an informal collegiate manner – wherein there may be greater opportunities and flexibility for getting a favorable resolution.

The following approach to dealing with these disputes may be of some assistance:

(i) Determine all the facts in relation to the dispute – preferably documented facts (particularly university regulations, guidelines), not hearsay – or opinion.

(ii) Organize an informal meeting with the other party to discuss the issues at hand.

(iii) Don't present ambit claims to a person who is a work colleague in the hope of negotiating them back later – describe sincerely and politely the requirements for resolving the dispute.
(iv) Verbally and courteously present the hard evidence supporting the case (formally presenting documents with rules and procedures may be viewed as inflammatory).
(v) If the other party is unwilling to shift from a previously stated position, politely ask if he/she can explain the reasoning behind his/her decision.
(vi) If the reasoning behind the decision-making is sound then there is little more that can be achieved, save to ask for advice on how both parties can work together to achieve a better outcome.
(vii) If the reasoning behind the decision-making is unsound or appears partisan or unreasonable, it may be appropriate to then up the ante by presenting documentary evidence that disputes the opposing view.
(viii) Ask the other party how he/she would proceed given the current circumstances.
(ix) If the total dialog has been unproductive or it appears that there are other *non-declared issues* at play, ask the other party about mechanisms for appealing his/her decision.

In the final analysis, one cannot expect to win every dispute through a solution that is entirely favorable. One has to accept that there are limitations to what can be achieved, and that each dispute is only a minor skirmish in the lifelong war for knowledge. It is important therefore not to exaggerate the stakes in a dispute or to make them personal but, rather to view them in the context of a professional working life that may go on for decades after the dispute is resolved.

9.6 Conflicts with External Partners

At some point in their careers, many supervisors will need to deal with external partner organizations as part of their research. These may include:

- Commercial.
- External research organizations.
- Other universities.
- Teaching hospitals.

These dealings bring with them numerous opportunities for conflict, particularly in relation to matters such as:

- IP.
- Shared use of facilities.
- Royalties.
- Joint or co-management of staff/students.
- Interpretation of legal agreements.

These sorts of disputes are in a different category to the others discussed in previous sections because, from the external party's perspective, they are engaging in a dispute with the entire university. In attempting to manage such conflicts it is therefore imperative that, at all times, consideration is given as to how:

- Any actions will affect the external perceptions of the entire university and goodwill or animosity towards it.
- Any claims or commitments will be reflected as being those of the entire university.
- Resolution of the dispute can allow the university to still be held in esteem by the other party.

The most important point here is that even though, at its core, a dispute may appear to be between individuals, the resolution needs to be seen to be as being between the university and the external organization. It is therefore imperative that such disputes are never allowed to become personal because whosoever speaks on behalf of the university carries the burden of the institution's reputation.

Even small institutions can have their reputation (brand) value measured in millions of dollars – world leading university brands are measured in billions of dollars of value. Every dispute with an external partner needs to be measured against this yardstick. So, a dispute with an external partner over a million dollars may seem significant, and worth winning, but perhaps not if it damages a billion dollar institutional reputation.

The key point here is that one needs to be extremely cautious in external disputes because there are very few officeholders of the institution who are normally authorized to make decisions that will impact upon the institution as a whole, and its reputation. As a general rule, an academic involved in a dispute with an external organization should never deign to speak on behalf of the university – or even declare the university's position – unless specifically authorized to do so – preferably in writing by a senior officeholder of the institution. To do so can have serious legal implications – and ones with binding significance.

9.6 Conflicts with External Partners

In general, a university only authorizes a small number of people to make decisions on its behalf, typically the Chief Executive Officer (CEO) or nominee (e.g., Vice Chancellor, Deputy Vice Chancellor, Chief Financial Officer, etc.). Often, the appointed decision-maker will take formal advice from the university's legal representatives, marketing department and external engagement office. This enables decisions to reflect a whole-of-institution approach that takes into consideration a broad range of factors.

It may be helpful for supervisors to consider the following approach when disputes arise with external partners:

(i) Ensure that, at all times, the dispute remains at a professional level and never engage in any activities or dialog that reflect personal antipathies with the external party.
(ii) Do not conduct formal dialog on the dispute with an external party via electronic or documentary correspondence unless the correspondence has been screened and formally approved by the university's legal representatives – also avoid any informal dialog via written correspondence, including electronic messaging, email, etc.
(iii) Determine and assemble all the facts regarding the dispute – preferably documented facts rather than opinions and hearsay.
(iv) Present the information to an authorized officeholder of the university and ask if an informal meeting can be held with the external partner to discuss the issues.
(v) If the external partner is willing to meet on an informal basis, then it may be opportune to use the meeting to ask the external partner to state its concerns and requirements for dispute resolution – no commitments should be given to the external partner other than to say that the matter will be referred to authorized officeholders of the institution for consideration.
(vi) Never state the university's position – even at an informal meeting – unless that position has been formally sanctioned by authorized officeholders of the university – preferably in writing.
(vii) If an authorized university officeholder has asked for the university's position to be conveyed to an external party at a meeting, then insist that that officeholder put the position/instructions in writing, so there is no possibility of a dispute over what has been called for. If the officeholder refuses to put in writing the position of the university, then send an email to that officeholder, echoing exactly what has been discussed, and ask them to acknowledge, by return email, whether or not this is the formal position of the institution.

(viii) If the external partner insists upon legal representation at a meeting, then such a meeting should not take place until authorized university officeholders have been advised and legal counsel sought from the university's legal department – if such a meeting proceeds, it should generally only be with the presence of a legal representative from the university.

The resolution to such conflicts can sometimes therefore be complex and based on far more than one representative of the university. However, if the supervisor is the person that has become sandwiched in a dispute as a result of his/her involvement in a collaborative project, then he/she becomes the face of the institution for so long as the dispute continues. Each discussion – formal or informal – therefore has to be viewed through the lens of the image that is being broadcast on behalf of the institution.

10

Research Supervision in Industry/Partnered Collaborative Research

10.1 Overview

It is increasingly common for postgraduate research programs to overflow the confines of an individual research group and spread into:

- Other university departments, laboratories or research centers.
- University teaching hospitals or research institutes.
- External research and development (R&D) organizations including government, military/defense/aerospace.
- Business and industry and government departments (including welfare agencies, social services, taxation, health, etc.).

A collaborative research program involving a postgraduate research student may emerge for any number of reasons, including:

- The postgraduate research program (or an over-arching research program, of which the postgraduate research is one component) is funded by an external partner.
- The research program is funded through a government collaborative grant which involves both the university and the partner organization.
- The research is dependent upon the use of equipment which is not available within the confines of a university research group (e.g., synchrotron).
- The research is dependent upon the use of human patients as subjects for experimentation, under competent professional supervision (e.g., in a hospital environment or clinic).
- The research is dependent on a range of different professional skills (e.g., engineering postgraduate research into an electronic biomedical implant device may require collaboration with a hospital surgical unit for implantation).

In any collaboration, there is an expectation that each partner within it will benefit in some way from the outcomes – for example, through a sharing of knowledge/intellectual property (IP) or through sales of a product or service arising from the research. Generally, this is defined in the context of some formal, documented agreement and it is unusual (risky) to have two partners (e.g., university plus commercial partner) collaborate until such a documented agreement has been formalized.

There are numerous benefits to collaboration, not the least of which is that a research student is exposed to a broader perspective than he/she might encounter within the confines of a small research group. If a collaboration takes place with business/industry, the collaborative research period can also constitute an internship, wherein a postgraduate researcher can become accustomed to commercial imperatives and limitations. So, the sum of the collaborative parts can end up being greater than the whole.

A graduate emerging from a collaborative postgraduate program can be significantly more:

- Street-smart
- Confident
- Erudite
- Employable

than one who solely completes a degree within a university research group.

However, with the advantages there also emerge additional challenges for both the research supervisor and for the research student, specifically in the areas of:

- Contracts and distribution of liability
- Joint management/supervision
- Definition of outputs/outcomes/deliverables
- Utilization/management of resources and funding
- Research student preparation
- Health and safety
- Adherence to multiple (potentially conflicting) sets of instructions and procedures
- Confidentiality
- IP
- Student welfare.

In summary, the most significant challenge with all these areas is that the research student ultimately becomes responsible to multiple masters, and there is potential for conflict, confusion and frustration. The supervisor's role is to ensure that these things do not occur in the first instance and, if they do occur, to resolve them as quickly as possible in the interests of the student.

In this chapter, we examine some of the key factors that need to be considered in order to maximize the opportunities for a successful collaboration.

10.2 Contracts and Distribution of Liability

Postgraduate research programs necessarily involve money, physical resources, personnel, obligations and liabilities. In the case of collaborative research programs with partner organizations, all these elements need to be documented carefully, and agreed to by each partner – preferably in writing – before the collaboration commences.

A collaboration agreement needs to cover a broad range of issues, including:

- The objectives of the collaboration.
- The commencement date, duration and expected termination date (and mechanisms) for the collaboration.
- The nominated representatives from the university and partner organization/s who will be responsible for oversight of the collaboration.
- The mechanism by which the project will be managed (e.g., management committee with representatives from partners).
- The financial (cash and in-kind) contributions to the project by each partner, and any specific resources to be devoted to the project.
- The mechanisms by which research staff (and specifically research students) can be allocated or appointed to the collaborative project.
- The timeline, critical pathways and milestone dates for the project.
- The specific outcomes/deliverables for the project.
- The distribution of IP and/or any royalties or benefits derived from the outcomes of the collaboration, including payment terms and/or schedule.
- The responsibilities for any liabilities during the collaboration.

Generally speaking, in collaborative research there can be many vagaries and unknowns associated with the process of discovery and, because any agreement is written prior to the commencement of research, it is not realistic to expect that the agreement will be an all encompassing document for the duration of the program. Nevertheless, it does represent the formal reference point from which disputes need to be resolved through negotiation, and a point of departure for ongoing agreements, as elements of the research begin to reveal themselves.

A written agreement/contract needs to be signed by authorized officeholders of the various partner organizations in order to reflect their imprimatur in the event of a dispute, and to carry organizational weight in terms of day to day project management. Unless a research supervisor is an authorized signatory for the university, he/she would typically require advice and approval from authorized officeholders of the university before any formal collaboration agreement could be signed with an external organization. This is especially true when such an agreement makes commitments of resources, staffing or funds on behalf of the institution, or when the agreement may lead to obligations or liabilities on the part of the institution – which is almost always.

The time has long passed when two organizations could simply enter into a verbal agreement on a joint program and work with goodwill towards achieving mutually beneficial outcomes. And, while many academics may consider the potential IP and royalties from a collaborative program as paramount in a collaboration agreement, in reality there are other factors which may dwarf these in terms of significance.

The reality of modern research is that the financial benefits accruing to universities from IP and royalties are generally small – with rare exceptions (e.g., development of a commercially successful pharmaceutical product). However, the potential liabilities to the university may be significantly higher.

Specifically, the following liabilities may need to be considered in any contractual arrangement between a university and partner organization:

- Personal injury arising to the research student or supervisor while working on the partner's premises.
- Personal injury arising to staff of the partner organization while working on the university campus.
- Penalties/sanctions arising from breach of confidentiality (e.g., on the part of the research student).

Academics often dismiss these sorts of issues as minor details – particularly when a research student is conducting research at a partner organization which appears to be relatively safe – for example, having only an office environment, rather than potentially hazardous machinery. The reality, however, is that a research student can be injured as severely by falling down on a slippery office floor, as he/she can be injured as a result of misadventure with dangerous machinery. Injuries invariably translate into significant costs and compensations, so it is necessary to understand which organization is responsible and/or insured if the worst should occur.

10.2 Contracts and Distribution of Liability

A research supervisor may make the assumption that a research student working at the premises of a partner organization will be covered by insurance held by the partner organization – and a partner organization may assume that its staff are covered by university insurance when on campus – neither of these assumptions are necessarily valid in practice. For example, it may be that staff from a partner organization working on a university campus may be considered as external contractors and may need to provide their own insurance. The same may be true for students in the partner organization.

Each partner to a collaboration therefore needs to determine, from the appropriate officeholder within their own organization, the conditions under which students or staff working in either the university or partner organization will be covered by insurance.

All these discussions in relation to collaboration contracts apply equally to entities which may superficially appear to be part of the university itself but may, in practice, be separate legal entities. For example, a teaching hospital with the same name as the university may, for legal purposes, have a completely different governance and insurance structure. It may also have its own cost centers and therefore may form a distinctly different entity for distribution of royalties or IP.

The following sequence of activities may be helpful in establishing collaboration agreements:

(i) Check university processes/procedures for establishing research collaborations and formal agreements.
(ii) Gather necessary information about the project and the partner entity – including project timelines, outputs, management team, resourcing/funding commitments, distribution of intellectual property and royalties, liabilities and obligations.
(iii) Work with the partner organization/s to get informal consensus on project details before expending time on written agreements.
(iv) Ensure that contracts are prepared by suitably qualified professional staff in either the partner organization or university.
(v) Liaise with university legal representatives and authorized officeholders to ensure that any contracts satisfy university requirements – work informally with the partner organization/s to iteratively achieve a document that is likely to be acceptable to signatories from all partners, including the university.
(vi) Ensure that any formal contracts are signed by official signatories of the partner organization and the university.

10.3 Joint Management/Supervision

10.3.1 General

Research that is undertaken in collaboration with an external partner – by definition – creates additional supervisors for the research student, whether this is formally instituted or not. Once a project leaves the confines of the supervisor's research group, and enters into a different premises or organization, one needs to accept that activities in that external area will come under the jurisdiction of outsiders.

It may be wishful thinking for a supervisor to believe they have sole charge of a collaborative project, but the practical reality is that they generally do not. In any collaboration, the needs of all the partners need to be met if the project is to succeed, and this is unlikely to occur unless the supervisor takes a team-based approach to management. Notwithstanding this need for teamwork, ultimately the supervisor can have practical responsibility for oversight of:

- The health, safety and general wellbeing of the research student while he/she is at the partner organization.
- Any university resources/funds that are utilized during the course of the collaboration.
- The health, safety and general wellbeing of any people, from the partner organization, working in the university.

10.3.2 Project Management Committee

Collaborative projects are highly dependent upon goodwill between the various partner organizations. Although it is generally the case that each collaboration will have a written contract that outlines timelines, outputs benefits and responsibilities, this only serves as a reference point for project management rather than an all-encompassing, prescriptive guide. As a collaborative project progresses, and elements which were unknown/unforeseen during the preparation of the contract emerge, they need to be managed – on a case by case basis – by representatives from the university and the partner organizations.

Collaborative research projects therefore tend to be managed by a committee, composed of representatives from each of the partner organizations. It is important that such committees are not so large that they become unwieldy but they do need to have sufficient members to facilitate the smooth running of the postgraduate research program and to remove any obstacles along the way. Consider Examples 10.1 and 10.2.

Example 10.1
A collaboration between a university and a manufacturing company may include engineers from the company to co-supervise and mentor the research student while on the company premises. It may also need to include a production manager or machine shop manager – who may have no direct oversight of the student's work, but will need to make prototypes and production modifications.

Example 10.2
A collaboration between a university and a teaching hospital may require a management team that includes relevant medical specialists from the hospital, and also representatives from critical supporting areas, such as pathology or perhaps medical imaging.

As a general rule, a management committee does not need to include staff who have no direct involvement in the technical aspects of the project. For example, it is probably unnecessary to have the company accountant present at meetings, simply because he/she manages the project finances from the partner perspective.

10.3.3 Supervision of the Research Candidate

Staff from partner organizations who are formally accredited, by the university, as supervisors for a research candidate working on a collaborative program can have a similar role to any other joint supervisors within the university system.

Sometimes, however, representatives from partner organizations in a collaboration may not have the necessary qualifications to act as a joint (or associate) postgraduate research supervisor, and a university may not be able to formally accredit them as such for the purposes of the project. Nevertheless, these partner representatives may still have valuable contributions to make to the student's postgraduate research program.

The university supervisor's role is to balance the input from non-accredited staff from partner organizations against the best interests of the research student in postgraduate research sense. This generally requires considerable skill.

It is all too easy for a supervisor to quantize the activities of research students into things that need to be done for the purposes of his/her research degree, and things which need to be done for the collaborative project, but which are not part of the research degree. In such a scenario, a supervisor might simply preclude a student from undertaking activities not related to the research degree. The danger with taking such a clinical approach is that it may alienate the partner organization to the extent where the entire project, including the postgraduate research elements, become unworkable. In order to understand the sorts of concessions and informal agreements that need to be made, consider Example 10.3.

Example 10.3
A research collaboration, involving a postgraduate research student, takes place between a university and an aerospace organization. The research student needs to undertake materials testing for his research, using facilities provided by the aerospace partner. A supervisor from the partner organization agrees to provide the facilities on the condition that the student undertakes some additional tests at the university, which have no direct relevance to the postgraduate research, but which do have benefit to the partner organization. What is the research supervisor to do?
If the supervisor flatly refuses to allow the student to work on tests which are not relevant to the postgraduate research, then that organization may withdraw facilities support for the tests the student does need to do because they are not critical to the outcomes of the project as they view them. This is where the supervisor needs to be able to tread carefully between the strict demands of the postgraduate research program, the individual-level demands of the partner organization, and the practical reality of bringing these demands together. There is also a point where a supervisor needs to differentiate and balance between goodwill, pragmatism and exploitation. For example, what happens if the collaborator wants ongoing work from the student which is not related to the postgraduate research? The supervisor needs to be able to draw the line in the sand using mature judgment. One approach might be to discuss the practical realities of such a situation privately with the research student and determine his/her perspectives before even forming a position to put to the partner organization.

Ultimately, a research supervisor has primary responsibility for the student and the conduct of his/her postgraduate research. This needs to be made clear from the outset. Without such delineation of responsibility, a research student is likely to receive instructions from multiple sources and this will create confusion and disillusionment. In the context of the collaboration, a supervisor needs to act as a filter between the various instructional sources and the student. A good approach is to ask the partner organization to run any requirements they have for the student through the supervisor in the first instance – before discussing them with the student. This issue is re-examined in Example 10.11.

In many collaborative research programs the research student may be invited as a member of the management team and may have a role to play in providing input to the project directions. This makes the role of the supervisor more complicated because the student becomes both a peer and an apprentice researcher at the same time.

Overall, however, the supervisor always needs to place at the forefront of his/her supervision the best interests of the research student – continuously finding the delicate balance between the theoretical requirements for the research and the practical realities of the collaborative environment. If these balances are explained to the research student then the student should benefit from the professional learning that comes from managing conflicting/contradictory objectives and competing interests. These are a practical reality of the professional world and both the supervisor and the student can benefit from dealing with these challenges.

10.4 Definition of Outcomes/Deliverables

In defining the outcomes/deliverables of a collaborative research program, it is important for research supervisors to keep in mind the old adage that, *in order to avoid disappointment, one should always promise less than one can deliver, and always deliver more than one promises.*

Over-promising is a key mistake in research collaborations, particularly on the part of a university, which has its reputation at stake when it makes such promises. In particular, there are differing perceptions of outcomes and their value – particularly between universities and commercial entities. For example, a university may perceive the testing of a hypothesis and the discovery that it is invalid as a useful piece of knowledge – a commercial partner may perceive this to be a failure, because it cannot lead to a business outcome. An industry partner may perceive a process or product, which is not well understood technically, but still yields commercial results as a success – a university may see this as a failure because it represents unsubstantiated science.

In a research collaboration, the partners need to be in synchronism when it comes to jointly defining research outcomes for a project. From the university perspective, and particularly in the case of research students, it is critical that a university understands its own limitations. Consider the Examples 10.4 and 10.5.

Example 10.4
"The objective of the research program is to develop a device which can be used to automatically determine the temperature of a machine tool bed, and which can be commercialized by the partner organization."

Example 10.5
"The objective of the research is to explore the development of a prototype device, and to determine if it can be used to automatically determine the temperature of a machine tool bed. If the prototype proves to be useful, the company will then need to undertake the development work required to generate a commercial product based upon the prototype."

Example 10.4 highlights a typical mistake in the definition of project outcomes on the part of a university. It suggests that:

- The research has a predetermined outcome which will be achieved.
- The research student can not only achieve the outcome but can build a product to a commercial standard.

This sort of mistake is also commonly made in projects where a postgraduate research project requires software development – and it is implied that a research student can deliver a piece of software to the sorts of standards required for commercial use – in general this is not the case.

In Example 10.5, on the other hand, the research is genuinely open-ended, and is not predisposed to achieving a particular outcome. Further, the research student is only required to develop a laboratory prototype model, and it is clear that the commercial partner has the responsibility of converting the prototype into commercial reality.

In trying to avoid the temptation of over-promising to an external partner organization, it may also be useful to divide the postgraduate research project into discrete elements, each with some tangible output that each partner can use to calibrate its oversight of the project. Table 10.1 provides an example set of milestone objectives.

The same commonsense rules should apply to defining project outcomes/deliverables as those that apply to developing a project management

10.4 Definition of Outcomes/Deliverables

Table 10.1 Sample project milestone/deliverables chart

Project Week	Milestone/Deliverable
5	Completed initial literature review with options for investigation
8	Project methodology defined and documented
16	Experimental design completed and ready for approval by project management committee
17	Experimentation commencement
25	Initial experimental results presented for evaluation by project management committee
32	Complete experimental results presented for evaluation by project management committee
39	Analysis of experimental results completed and documented
42	Documented recommendations for development of prototype presented to project management committee
⋮	⋮
156	Handover of prototype and evaluation data for development and commercialization to commence

chart – that is, don't waste time on defining outcomes/deliverables unless one is fully committed to achieving them. In the case of postgraduate research, it is the research student that is responsible for delivering outcomes and it is therefore important that he/she is genuinely confident of delivering them.

In most collaborative research projects, a management team will create a timeline of milestones/outcomes/deliverables as part of the project development – and before any significant research work has been completed. Unfortunately, this is often presented to a recruited research student *fait accompli* at the beginning of the candidature – and it is assumed that the student will be able to comply with the stated plan. There are problems with this.

Firstly, the student has to genuinely buy-in to the plan and the deliverables, rather than to merely have them handed to him/her. Secondly, students who are recruited to the project may agree to the plan merely in order to please the management committee and without a firm belief in being able to achieve the objectives. A better approach may be for the project management committee to present the initial plan to the research student as a draft – and to have the research student come up with a final plan for consideration by the committee. This provides a higher level of buy-in to the project on the part of the student.

As with all project management fundamentals, all parties need to genuinely commit in order to get outcomes. This means treating the milestone dates and outcomes as firm and immovable. Specifically, this will involve:

- The project management committee ensuring that resources required for the completion of the research are provided to the student as initially agreed and on time.
- A research student increasing the number of hours devoted to the project each week in order to meet deadlines – rather than just pushing back deadlines.
- The project management committee providing additional resources if it becomes apparent that the research student is working at maximum capacity and due dates are at risk of not being met.

10.5 Utilization/Management of Resources and Funding

10.5.1 Fundamentals

A collaborative research project will generally involve people, funding and equipment which, ultimately, all come down to funding. A collaborative research project may be funded by:

- A government or benefactorial grant specifically directed towards a collaborative research project involving the partner organizations.
- Joint contributions to a total funding pool by the partner organizations.
- A university (in-house) grant to a research supervisor to facilitate a collaboration with external partners.

A written collaboration agreement needs to pre-date the commencement of the actual research project and define which expenditures of resources are to take place and when. Further, the agreement needs to define who will be the authorized signatory (or signatories) for expenditure of funds. This may be the university or one of the partners – or a joint management committee and/or its nominated representative (e.g., project director).

In any such project there will be regular, recurrent expenditures (e.g., for scholarships and salaries), as well as occasional expenditures on oneoff items (e.g., purchase of equipment or services). Careful budgeting and book-keeping is particularly important, as is the monitoring of cash flow. In particular, it should not just be assumed that because a university is a large entity that it has the capacity to release any amount of funding at any time. A large collaboration may involve the expenditure of millions of dollars, for infrastructure or equipment, and it should not be assumed that the university automatically has the cash-flow to cover such expenditure at a moment's notice. Large expenditures within a university generally need to be scheduled into the organization's financial management systems in order to avoid cash-flow problems.

A government grant to a university may, in practice, be paid in installments (e.g., monthly, quarterly or annually). The project management committee (or its nominated representative) needs to ensure that recurrent expenditures do not exceed the account balance at any given time and, if there is a chance that they might, then the organization holding the funding needs to be notified to determine if accounts can go into the "red". Consider Example 10.6.

Example 10.6
A basic collaboration is funded by a government grant which is paid quarterly. However, the project involves the purchase of large equipment and the payment of a scholarship. An initial purchase of equipment may send the project account balance into the red, and leave insufficient funding account to pay for the scholarship until the next installment arrives. Managing cash flow issues is therefore important and, if it is to be the university that is the holder of the account, then the research supervisor needs to liaise with his/her line manager to ensure that the institution can facilitate funding for a scholarship while an account is in the red.

10.5.2 Accountability/Auditing

Universities tend to be large organizations that, by necessity, use sophisticated accounting and auditing systems to monitor income and expenditure. A large institution may have many thousands of accounts and cost-centers that need to be managed, and each has nominated signatories.

Within an institution the basic rules relating to expenditure of funding are relatively straightforward and based upon common sense:

- Funding from a project account is generally only expended according to a pre-determined budget.
- Auditing is typically carried out against the budget.
- Signatories to accounts can generally only expend within budget and, even within those constraints, are often limited to a ceiling amount on any one expenditure item – unless they have approval from a line manager or have undertaken a regulated purchasing program (e.g., competitive tendering).
- Signatories to an account cannot allocate/expend funding to themselves – either directly or indirectly – without the authorization of a line manager who is not accountable to the signatory (i.e., a signatory to a collaborative project account may not sign for a payment or reimbursement to himself/herself – this would need to be approved by a more senior person).

190 *Research Supervision in Industry/Partnered Collaborative Research*

- Significant expenditures need to be justified in terms of supplier (e.g., has a competitive tender process been implemented or have multiple quotes been received prior to purchasing an item?).
- Some expenditures can only be made through particular suppliers (e.g., computer hardware, stationery or airline travel may need to be purchased through a university-certified provider).
- Signatories to an account need to formally disclose any conflicts of interest or pecuniary interests that may impact upon expenditure (e.g., an account signatory cannot sign funds to an external company of which he/she is a major shareholder – or a spouse or relative is a major shareholder – unless this has been declared – and the expenditure approved by a line manager).

Modern accounting systems require payments from an account to be made via expenditure codes – the expenditure codes are generally defined by the organization setting up the system, and are peculiar to the types of expenditures that that organization would make. For example, a university might have expenditure codes for stationery, equipment repairs, etc. Further, it is often the case that an institution requires any suppliers that it deals with to have a formal, documented supplier relationship with the institution – in other words, the accounting system needs to recognize any organization to which payments are made.

These checks and balances are designed so that auditors can ensure that institutional funds are expended against items that make sense in relation to a budget, and that any external organizations to which payments are made are registered with the institution and can be investigated (e.g., shareholdings identified) to determine whether any conflict of interest irregularities are taking place.

All these checks and balances mean that if it is the research supervisor who is responsible for the project budget, then he/she needs to exercise particular care with what is expended, and which organizations receive payments. Irregularities and anomalies will be flagged, and the signatory may be required to explain them. Consider Example 10.7.

Example 10.7
A supervisor has used research project funds to purchase a dishwasher from a department store – for cleaning project glassware. This might be flagged as an irregularity by a university audit, because it would not be something normally purchased for a research project. In order to ensure that such an unusual purchase was not acquired for the supervisor's personal benefit, an

audit team may require the supervisor to substantiate why such a purchase was made with research project funds, and why a particular supplier was chosen.

10.5.3 Resources

There is a common perception by academics that institutional resources are, by default, free of charge, merely by virtue of the fact that they exist within the institution and are available for use. This isn't the reality in many universities, and it is certainly not the case in commercial enterprise. Most organizations operate some form of cost-centered accounting, wherein capital equipment, consumables, technical support, etc. need to be formally considered in budgeting.

It can be the case that within an individual department/faculty or research group, some infrastructure is budgeted centrally, and those resources are made available to staff as part of their working environment. Effectively, this means that each member of staff has had part of their budget allocations diminished to fund this central activity. However, this sort of infrastructure socialization rarely applies across departments, faculties or operational units.

By way of example, consider that a university hospital may have a medical imaging device which is required for a postgraduate research program. Usage of such a device is likely be costed to each user (per image or per time unit) as would any complementary technical support.

It cannot be therefore be assumed in the context of a research collaboration that resources which are available within the university are resources which are free to include within a research collaboration. The same basic practice is all the more rigorously administered within a commercial environment, where resources and supporting services are considered as a cost or profit center within the organization.

Many universities around the world are government funded, and so it is altogether common for commercial partners to assume that – as taxpayers – their entity can make use of university resources and supporting services at no cost. It may be the case that, in order to facilitate a collaboration, the university contributes resources and supporting technical services *gratis* as its in-kind contribution to the project – in the expectation that it will receive returns at some later date, in the form of IP or royalties. The exact nature of what resources can and cannot be provided needs to be discussed with authorized university officeholders prior to making any commitments to partners.

It is also necessary to determine what resources a collaborating partner can contribute to a project as in-kind, and what resources and supporting services can only be supplied as costed items, which need to be included within the project budget.

From an auditing/budgeting perspective, extreme care needs to be taken with allocating project funds to a collaborating organization in exchange for resources or services provided back to the university. Consider Examples 10.8, 10.9 and 10.10.

Example 10.8
A government grant is awarded to a university to engage in a collaboration with a company. The company and the university both contribute cash and in-kind resources to the project. However, excluded from the in-kind contributions from the company are tooling and workshop labor for generation of a prototype – these are to be billed to the project budget by the company.

Example 10.9
A government grant is awarded to a university to engage in a collaboration with a company. The company and the university both contribute cash and in-kind resources to the project. One of the project requirements is to purchase a piece of software valued at $500,000. There are several suppliers for the software but the company insists on the project using its own software which has been developed in-house for the project. $500,000 would then be transferred from the project budget to the company.

Example 10.10
A government grant is awarded to a university to engage in a collaboration with a company. The company and the university both contribute cash and in-kind resources to the project. Part of the university's contribution to the project is three postgraduate research scholarships. The research supervisor wants to be paid a consulting fee by the company because he feels that his expertise, over and above his supervisory work, will be required to facilitate completion of the project.

In Example 10.8, a small proportion of the project budget is deployed to providing fabrication work for the research. It may be that the tool-making department within the collaborating company is a separate cost center outside the control of the collaborating staff and therefore this could be a legitimate expense.

10.5 Utilization/Management of Resources and Funding

Example 10.9 is fraught with problems and may cause serious questions to be raised under audit. In this example, a piece of commercial software has been valued at $500,000. This may be the legitimate value for the software when sold in the commercial marketplace. However, the incremental cost to the company providing it to the project is negligible (i.e., only the cost of the medium or network transfer time). In effect, the company is receiving a direct payment of $500,000 from a project budget for something which has negligible incremental cost. In this example, the university should have negotiated the provision of the software at no cost to the project during the formulation of the agreement.

Example 10.10 would also cause serious problems under audit and constitutes a serious ethical problem for the supervisor. A university is contributing considerable cash to a project in the form of scholarships – from which the company directly benefits as a result of the research labor provided. The company has been asked to pay funds to the supervisor to support the inadequate knowledge base of the research students. In effect, in this scenario, it can appear that the supervisor is giving away university resources (i.e., research scholarships) in exchange for personal financial benefits (i.e., consulting fees) from the partner organization. This constitutes a serious conflict of interest, and is not something which an academic supervisor should even contemplate (much less agree to) without full disclosure of personal benefits to his/her line manager. Even then, such an arrangement could leave the academic open to charges of serious misconduct, in the sense that he/she has traded university resources to a commercial partner in exchange for personal financial profit from that partner.

The following points may be of assistance in managing resources related to collaborative postgraduate research programs:

- Do not assume that resources (infrastructure, laboratory equipment, technical support, etc.) either in the university or the partner organization are available for use free of charge – generally, even in-house resource usage incurs costs from other departments.
- Determine all resource costs and availabilities in both the university and the partner organization before formulating a collaboration agreement.
- Seek advice from authorized officeholders prior to committing any university resources to a project.
- Exercise extreme caution in paying for resources/services from collaborative research project budgets – conflicts of interest and pecuniary interests need to be identified and disclosed, before project funds are approved for expenditure.

- Residual resources from a collaborative project either need to be disbursed as per the collaboration agreement or, if no such consideration has been given in the agreement, advice from appropriate university officeholders needs to be sought on any distributions.

10.6 Preparation of Research Student for Collaborative Research

10.6.1 Overview

There are considerable benefits accruing to a postgraduate research student who elects to undertake his/her research as part of a collaborative research effort. A collaborative research project can provide:

- Broader exposure to the field in both an academic and commercial sense.
- Possible commercialization opportunities.
- Opportunities for the student to develop negotiation skills.
- Opportunities for the research student to pursue a career outside the university – with the partner organization.

However, there are also considerable challenges for the student, and it is the supervisor's role to ensure that these do not become insurmountable.

It is neither sufficient nor responsible for a supervisor to simply place a student in an outside environment without consideration of the skill set that the student will need in order to function in that environment. It is also increasingly common for students from various international backgrounds to be placed in partner organization as part of collaborations, and these students can experience even higher levels of difficulty and anxiety than local students. Specifically then, a supervisor should consider the following points and determine how best to brief the research student accordingly:

- Basic induction into the company – health and safety
- Dress codes
- Dealing with company staff
- Organizing meetings with company staff
- Confidentiality
- IP
- Bullying and harassment.

It is a mistake to assume that because a research student has a high level of intellectual capacity in a scholastic sense that they have the business and commercial acumen (i.e., *street smarts*) to make their own way in a partner organization.

10.6 Preparation of Research Student for Collaborative Research 195

The time expended on ensuring that a student is suitably briefed for his/her role in a collaboration is more than recovered by the reduction in the number of potential problems that would otherwise arise without briefings.

10.6.2 Basic Induction into the Company – Health and Safety

It should not be assumed that a student has had sufficient exposure to environments outside the learning environment such that they will automatically be able to fit in to a partner organization. Further, it may be the case that, as part of a collaborative agreement, the research student has no formal position or authority within the partner organization. This makes the research student an outsider to the partner organization and it makes his/her role all the more difficult.

There are many basic pieces of information which would be given automatically to a normal employee in an organization – often these are not supplied to the research student, so it is important that the research supervisor ensures that they are covered. Specifically:

- Basic access to the organization – keys; smart cards; car-parking arrangements.
- Rudimentary resources – photocopying; access to IT/networking; telephony; stationery.
- Tour of the organization – including location of relevant facilities and accessibility; location of emergency/fire exits.
- Explanation of day to day processes – meals, refreshments, office activities, etc.

These may appear to be trivial issues but they are not trivial to a student who has never worked in a commercial environment, and they are particularly daunting for students from international backgrounds who are not used to local customs.

In addition to these basic requirements, however, there is also a need for the organization to formally induct the research student into any local health and safety practices that need to be followed. This is particularly important where the external partner organization is involved in activities such as:

- Laboratory work
- Manufacturing
- Chemical/biological processes
- Mining
- Building and construction.

It is also important that basic briefings are given where an organization is involved with particular types of clients, particularly in medical/social work situations (e.g., patients with Alzheimer's, frail elderly people, patients with contagious diseases, etc.).

The research supervisor has a responsibility to ensure that the partner organization formally undertakes health and safety training before a research student commences activities there. Further, a research supervisor should meet with the research student to ensure that any such training which has been received is adequate and, if not, determine how any shortcomings can be addressed.

10.6.3 Dress Codes

One of the most overlooked aspects of briefing research students in relation to deportment in a collaboration relates to dress codes. This may appear to be a trivial issue but it has significant consequences. A research student working in a collaborative project needs to be able to win over staff in the partner organization for ongoing support, and inappropriate dress may cause immediate alienation, which is difficult to overcome.

An organization may have an implicit or explicit dress code but, in either event, respect for the code is a sign of respect to the staff working in that organization.

During the formulation of the project, a research supervisor should acquire a good perspective on what sort of dress code will be required from the research student. It may also be useful to raise this with staff in the organization who will be collaborating with the university. Failing all the above, the student needs to be advised to develop an awareness for what is and is not appropriate in terms of attire.

Typically, a business, accounting, consulting or legal organization may have rigorous dress standards with formal business attire required. It would therefore be inappropriate for a student to arrive at work in such an environment with casual clothing. Conversely, an information technology company may have an informal dress code, and it would be equally inappropriate for a student to work in such an organization with formal business attire.

10.6.4 Dealing with Company Staff

Many research students may not have experienced the professional, commercial work environment prior to the commencement of their candidature. While a collaborative research project may give them the opportunity to gain

10.6 Preparation of Research Student for Collaborative Research

this experience, it is also important that their inexperience does not become an unnecessary liability to a collaborative project. In particular, research students are in a difficult position because the staff with which they will liaise may have numerous other responsibilities, which are more pressing on a day-to-day basis than the research project. The research supervisor needs to advise a research student:

- *How to address staff in the partner organization* – some staff may object to being addressed by their first name, others may see this as acceptable.
- *When to contact staff and when not to contact staff* – staff are busy and should not be bothered with trivial issues, or with issues that they do not have the capacity to resolve – in some organizations, office politics are in play, and students also need to determine which staff should not be contacted in relation to the project.

In addition, it is important that research students are not continually pestering staff in the collaborating organization with problems – they need to develop efficient communications skills wherein they can:

- Determine which staff member is best suited (and has the authority) to resolve an issue.
- Present the issue to the staff member in a short, concise manner.
- Present a potentially feasible solution to the issue, which the staff member may then be able to approve or facilitate.

It is important that the research students are seen as contributors, who are moving things forward, rather than complainers and time-wasters that are constantly highlighting shortcomings without presenting a proposal for action.

10.6.5 Organizing Meetings with Company Staff

Meetings can contribute to lost productivity in commercial organizations, so it is important that research students are not seen to be the cause of unnecessary meetings, or meetings which have no clear purpose.

Research supervisors need to ensure that students are fully briefed on how and when to call meetings that involve time from staff in a collaborating organization. Specifically, the meetings need to have:

- Clear objectives, to which those attending can contribute specifically.
- Attendees who are able to facilitate actions, or need to be briefed on developments and potential actions – meetings should not be overloaded with staff who only have peripheral relevance to activities.

- A concise agenda, from which actions can be derived and responsibilities for those actions allocated.
- Progress reporting on the status of previously allocated actions.

In some instances, the supervisor will need to be present at the meetings and in other instances, where discussions are more focused on issues internal to the partner organization, the supervisor may not be present. It is therefore imperative that the student is able to manage himself/herself professionally in those meetings – and that staff in the collaborating organization see them as productive.

10.6.6 Confidentiality

In collaborative research programs there are generally confidentiality provisions in place in regard to the project; its outcomes, and intellectual property. It is often the case that a research student electing to opt in to a collaborative project will need to sign some form of confidentiality agreement.

Any confidentiality agreement signed by a student will have consequences – some of these may impinge upon the student's ability to carry on a professional career in a specific area at the end of the research project. It is important that the research supervisor advises a research student to seek independent, professional advice before signing any such agreement. This is discussed further in Section 10.8.

10.6.7 Intellectual Property (IP)

A formal research collaboration with two or more partners will generally lead to an agreement on how to distribute/allocate IP and/or royalty income. A research student, opting into such a project and being a key contributor to the IP, may need to be a formal party to such an agreement.

As with a confidentiality agreement, an IP agreement is a legal document which may have implications for the research student's future career or aspirations to start his/her own business enterprise. It is therefore important that the research supervisor advises the student to seek independent, professional advice before signing any IP agreements. This is discussed further in Section 10.10.

10.6.8 Bullying and Harassment

A research student, conducting his/her research in a partner organization, can be subject to the rules and regulations of that organization, in addition to those within the university itself. Sometimes, there will be anomalies between processes and procedures within the university and the partner organization. However, when it comes to issues such as bullying and harassment, it is particularly important for the supervisor to advise the student that he/she is available at any time to discuss concerns that may arise within the partner organization.

From the supervisor's perspective, the logical position is to adopt the more rigorous of the processes and procedures of either the university or the partner organization as a reference point. This is particularly true if a student complains of bullying or harassment within the partner organization. It also needs to be considered that a research student may not technically be a staff member within the partner organization, and that some of its rules will not be applicable.

The supervisor should become the first port of call for the student when any issues arise in relation to bullying and harassment – or even if the student is accused of bullying and harassment.

10.7 Health and Safety

In any research supervision and in any research collaboration, the health and safety of the research student is clearly paramount, and supersedes any other considerations relating to the project.

In general, universities around the world tend to have rigorous health and safety programs for staff and students, but there are some necessary exceptions. It is often the case that universities have dispensations from national or regional laws, simply because it would otherwise be impossible to carry on laboratory or research work. For example, an electrical laboratory may be exempted from a requirement to have all electrical terminals insulated and for no bare wiring to be exposed – otherwise it would not be practical to wire up power circuits. In exchange for such allowances, universities often expend considerable effort to ensure that alternative safety measures are in place (e.g., earth leakage and additional current sensors in electrical laboratories).

The point here is that research students may have become accustomed to the unique health and safety considerations of the university environment, and subsequently need to work in a partner organization which is subject to more conventional regulations. The partner organization may have health and safety requirements which are more rigorous than those in the university but with fewer special-purpose safety fallback measures.

The research supervisor needs to ensure that the student is:

- Fully aware of the health and safety procedures in the collaborating partner organization.
- Briefed on the differences between conducting work in the university environment and in the partner organization.
- Provided with an appropriate health and safety induction program in the partner organization.

As with most aspects of the collaboration, ultimately the supervisor needs to bear responsibility for the student's welfare on the campus and on the premises of the partner organization. The supervisor therefore needs to ensure that he/she is satisfied that the partner organization is a sufficiently safe environment in which the student can work. Consequently, it may also be determined that the partner organization is *not* a sufficiently safe place for a student to conduct research. For example, the partner organization may be a small start-up company which does not yet have fully developed health and safety protocols.

It may also be the case that the partner organization is equipped for health and safety in its primary area of interest but is branching out into a new area through the collaborative research program – and that area is not serviced adequately by health and safety procedures. All these things need to be taken into consideration from the outset of the collaboration. If there is any doubt about the partner organization's capacity to provide a work environment that meets national/regional health and safety requirements, then the research student should not be permitted to work therein.

A research supervisor should also insist that the research student contact him/her if there are any concerns about any health and safety issues within the partner organization. The research supervisor has a moral (and often legal) obligation to act on these as a matter of urgency.

10.8 Adherence to Multiple (Potentially Conflicting) Instructions and Procedures

A research student in a collaborative research program can become the principal interface person between the university and the collaborating partner. Needless to say, this means that the research student will need to deal with two different work environments, often with contradictory or conflicting objectives. For example, a commercial organization may have:

10.8 Adherence to Multiple Instructions and Procedures

- Short timelines
- Day-to-day imperatives
- A focus on confidentiality.

A university, on the other hand:

- Works on long-term timelines
- Has few day-to-day imperatives
- Has a focus on publishing knowledge which is uncovered.

Learning how to balance and satisfy seemingly contradictory objectives is an excellent learning vehicle for the research student, and provides invaluable training beyond the basic research skills which are acquired. However, from the supervisor's perspective, it is important that the student is not faced with insurmountable challenges and contradictions – or, perhaps, in attempting to satisfy contradictory requirements, ends up working two jobs.

Table 10.2 highlights some of the contradictions that can exist between the university environment and the commercial environment. Supervisors and research students both need to be aware of these, and consider how they can bridge gaps when they arise.

The research supervisor and the research student also both need to be acutely aware that, when working in a collaboration, it is not reasonable to always draw hard boundaries between academic requirements and the requirements of the partner – particularly if the partner is a commercial organization. A collaboration requires give and take on both sides and, without this, the likelihood of any successful outcome is diminished significantly.

A collaboration agreement, established between a university and a partner organization, only covers the basic elements of what will be required for the completion of the project. Beyond these basic elements, it will often be necessary to negotiate additional items, and this cannot be done effectively unless there is goodwill between the two partners.

Establishing goodwill with a partner organization will require flexibility from the supervisor and the student. Consider Example 10.11, which poses a similar challenge to that highlighted earlier in Example 10.3.

Table 10.2 Differences between academic and commercial imperatives

	Commercial Perspectives	Academic Perspectives
(i)	Commercial/financial success	Individual academic excellence
(ii)	Converting ideas into functional products or systems is paramount	Testing the limitations of ideas is paramount – the objective is to see whether or not ideas can work

(Continued)

Table 10.2 Continued

	Commercial Perspectives	Academic Perspectives
(iii)	Ideas that cannot be converted into functional products or outcomes are a failure	Ideas that cannot be converted into functional products or outcomes are an integral part of the research process – the research is to assess boundaries not to achieve a perceived "correct" solution
(iv)	Relative competitive advantage – shades of grey	Absolute solutions – black and white
(v)	Corporate excellence – individuals collectively work towards achieving an outcome for the corporate entity	Individual Excellence – individuals work towards achieving an outcome for the individual
(vi)	Team-based projects with knowledge vested over a range of individuals	Individual project with knowledge vested in one individual
(vii)	Propensity to keep developments confidential	Propensity to publish any new developments as a hallmark of success
(viii)	Projects are multi-disciplinary in nature and may involve marketing, science, production, sales, etc.	Projects are highly specialized in nature and focus on one particular type of expertise
(ix)	Financial indicators are used to measure success	Academic rigor takes precedence over financial considerations
(x)	Projects are successful if they can be rapidly commercialized	Projects are successful if they contribute new knowledge
(xi)	Concepts have little value relative to the overall process of development and commercialization	Concepts are the end objective
(xii)	An academic solution is only a concept and a starting point	An academic solution is the end-point of a research process
(xiii)	Commercial outcomes take precedence over process rigor	Process rigor takes precedence over commercial issues
(xiv)	Time-frames are backward scheduled from perceived market demands for products	Time-frames are forward scheduled from an original concept
(xv)	Professional time-frames are measured in days or weeks	Professional time-frames are measured in months or years

Example 10.11
A research collaboration takes place between a materials science department in a university and an automotive component manufacturer. Unforeseen during the formulation of the project, it becomes apparent that the research student will need to conduct additional experiments for her research – and these

10.8 Adherence to Multiple Instructions and Procedures

will require unbudgeted, additional expenditure from the manufacturer, for machine shop work and materials. The manufacturer has some materials testing which needs to be carried out on university equipment, and the university has agreed to allow usage of its machine, provided that the research student agrees to do the work. This testing, however, is not in any way related to the research student's research. What is the research student to do?

One obvious solution is for the research student to simply conduct the unrelated experiments as part of a *quid pro quo* arrangement, provided that these are not too onerous. However, what is the supervisor to do if he/she acquiesces to this request and subsequently the collaborative partner makes another and another?

It takes skill and maturity to manage these sorts of issues carefully, and without offending the partner organization. A good approach might be to acquiesce to the first request and simultaneously advise the partner organization that this is only a one-off, goodwill gesture made by the university – in gratitude for the additional contributions of the partner, and to demonstrate commitment to the project.

There may be other challenges as well. For example, a research student may simply refuse to perform any work which does not form part of the postgraduate degree. This is the right of the student, but the supervisor's role is to then explain the consequences of that decision in the context of continuing the project. For example,

- What if the student subsequently requires additional support at a later time from the partner organization?
- What if the student is applying for a job after postgraduate completion and the organization to which he/she is applying asks for a reference from the collaborating partner?

These are all things which should be considered.

In a more general sense, there are also issues with the postgraduate research student receiving instructions from multiple sources, and having to interpret which ones to follow. A good approach is to have regular meetings of a management committee, composed of the supervisor and partner representatives, to ensure that there is consistency of message, and that any conflicting messages can be ironed out at the meeting. But what about day to day conflicts? Whose instructions should the research student follow?

One answer is to only follow those instructions of the supervisor, but this isn't always practical. Suppose that the supervisor requests for tests to be carried out at a partner organization but that organization advises the student that the relevant machinery is currently unsafe? There have to be basic guidelines that the research student follows independently of who issues instructions. For example:

- The research student will always adhere to health and safety instructions from either the partner organization or the university, and whichever of those entities precludes an activity because of health or safety concerns has effective veto power on the activity.
- The research student will follow the instructions of the collaborative project management committee, as a first priority, and instructions from the university research supervisor as a second priority.
- If a partner organization issues instructions and they are reasonable of themselves, and the student is able to follow them without impinging on the project, then he/she is at liberty to decide how to follow them – otherwise, the instructions are referred back to the university supervisor.

All these things may appear self-evident to a senior professional, or even one who has worked in the professional environment for some time, but they cannot be assumed to be skills possessed by a recent graduate. The supervisor needs to assess, based upon the student's professional experience, exactly how much advice needs to be given to enable that student to cope with multiple environments and multiple perceived bosses.

10.9 Confidentiality

Universities celebrate discoveries of knowledge and are quick to publish them for the benefit of the broader community, but this sort of freedom is generally curtailed as a result of a research collaboration – particularly if one of the research collaborators is a commercial organization. It can also be the case that, even when the partners are all universities or research organizations, collaborations take place so that the partners can derive a mutual benefit from any arising IP.

In addition to outcomes from the collaboration itself, there are other confidentiality issues that need to be addressed. For example, a commercial partner may allow academic staff and research students from a collaborating university access to processes, services, designs or software coding which that organization views as highly confidential – and which would otherwise not be made available to outside organizations. Release of such information could damage the financial position of the commercial partner.

Moreover, it isn't simply technical processes and designs that may be the subject of confidentiality. Consider Example 10.12.

Example 10.12
An automotive manufacturer is a collaborative research partner and is involved in litigation over the safety of one of its vehicle braking systems. As part of a collaborative project, on the design of a new braking system, the university and research student are given access to a range of information that impinges on the current litigation. This information is not only technical, but also includes correspondence that has taken place with customers – its release could be prejudicial to the automotive company in any litigation.

Clearly, extreme care needs to be taken with any commercially sensitive material which is revealed to the university or student as part of a research project. A collaboration agreement should therefore enunciate any specific conditions relating to confidentiality in respect of:

- Emerging IP/discoveries/research outcomes.
- Services, process, designs and software that are proprietary to any of the partners.
- Confidential (privileged) information which will be disclosed to the collaborating partners.
- Publication/publicity in relation to the research project and who can authorize its release.

In the case of Example 10.12 it would not be realistic for the university or research student to expect to publish work that contains information which impinges on a litigation process – and the university and the student need to understand that from the outset. More broadly, it may be agreed that all partners to a collaboration will desist from publication of any information which will have a negative impact on any one partner – without that partner's express approval.

The confidentiality issues are therefore serious ones, and they obviously impinge on the natural function of a university to conduct research for the purposes of knowledge discovery, and dissemination of findings for the benefit of society.

In the context of the moral dilemma of presenting a complete and accurate depiction of reality in collaborative research, part of this problem is resolved by making a statement of the environment in which the research was conducted. Consider Example 10.13.

Example 10.13
The research documented herein was conducted as part of a research collaboration with the Xylon Pharmaceutical Company, which contributed resources and funds to this project.

Example 10.13 provides a mechanism for disclosure of potential bias by virtue of environment. That is, it is a statement to the recipient/reader of any work produced that any outcomes need to be read in the context of the environment in which the work was conducted. In other words, that the research is necessarily constrained in its scope and outlook. Within that constrained environment, the work which is presented still must have complete integrity. Regardless of any constraints imposed by a partner, an academic or student should never publish material which he/she knows to be misleading to the reader, or an inaccurate depiction of the totality of the knowledge at hand. This includes misleading a reader by omission of critical information.

If restricted/confidential information is critical to an accurate picture of reality then clearly research cannot be published without it – and therefore, in all likelihood, the research should not be published at all.

In addition to constraints on making certain information public, confidentiality agreements may also preclude academic staff (e.g., supervisors) and research students from using any of knowledge derived during the course of the collaboration for later research work

The bottom line is that those who are involved in negotiating a research collaboration need to be mindful of the consequences of the confidentiality provisions of a collaboration agreement. Specifically:

- The morality of keeping confidential research findings which have a negative impact on one of the partners but are otherwise of significant consequence to society (e.g., the discovery that a product or process produced by a partner organization has a deleterious effect on individuals or the environment – or perhaps that a pharmaceutical product has no beneficial effects).
- The impact of confidentiality on the current and future careers of those involved in the collaboration, including academic staff and research students – particularly as it relates to the use of knowledge they have created in future research or career choices.
- The impact on academic performance metrics of not being able to publish research work which might otherwise be published.

The key point here is that these issues need to be contemplated *prior* to the signing of any agreement, and not the subject of a dispute after the agreement has been signed. If the university, as a stakeholder in a collaboration, feels that it is unable to live with the various permutations of confidentiality provisions, then it may be best that they do not enter into a collaboration in the first instance.

A research collaboration can bring about major benefits for the university, staff and research students but every benefit has a cost that needs to be considered, and confidentiality is a significant one of those costs.

10.10 Intellectual Property (IP)

The potential IP arising from a research collaboration needs to be identified prior to the commencement of the program, and arrangements made for its allocation at the end of the program.

A discussion on IP needs to be prefaced by noting that only a very small proportion of the intellectual property owned by a university generates significant income. Often, a university will have only a few major IP success stories – over decades of research – and these swamp much of the remaining holdings, which only create small income streams. It is important therefore not to become overly preoccupied with the retention or ownership of IP, given that the chances of achieving significant financial success with it are limited.

It has already been noted herein that a common *rule-of-thumb* is that for every dollar invested in research, ten dollars need to be invested in development and a hundred in commercialization in order to make a saleable/marketable end-product. At best then, the university is likely to receive less than one percent of a successful product revenues emanating from its research, and only a minute percentage of research will lead to a product – and, even then, only a minute percentage of the products will be successful products. These sorts of statistics need to be kept in mind before unnecessary efforts and funds are expended in negotiating IP agreements over what may amount to negligible monies. Consider Examples 10.14 and 10.15.

Example 10.14
A university enters into a research collaboration with a high precision machine tool manufacturer. The objective of the exercise is to investigate the materials science associated with making the machine beds in order to make the beds more thermally stable, and thereby improve machine tool accuracy.

Example 10.15
A university enters into a research collaboration with a pharmaceutical company to investigate variations to hypertensive drugs, in order to create a more efficacious product with fewer side-effects.

In Example 10.14, the university is engaging in research that is potentially contributing to improvements to only one part of an end-product (i.e., the machine tool). The research is looking at modifications to materials science that is already in place – and the proprietary knowledge of the commercial partner. The machine tool is a specialist professional device and not a general-purpose consumer product, so volumes are low. The chances of receiving significant returns from the IP – as commercialized by the partner company – even if the research proves to be successful, are limited.

Conversely, it may be the case that the knowledge derived from the materials science investigation has widespread and lucrative commercial applications in other areas not serviced by the machine tool company. The decisions relating to distribution of IP in this example therefore require considerable thought. Signing away the IP, in exchange for royalties or a direct payment, may provide a short term income but may preclude the university from using its own knowledge for other, more lucrative research in the future. Being excessively protective of the IP may be costly in terms of legal arrangements, and the returns may be well below the cost of the legalities.

In Example 10.15, the university and the partner company are jointly working on what is essentially a new product. The product is aimed at a mass consumer market; the cost of the product is significant, and the production volumes are large – so too are the costs of commercialization and approval by national regulatory agencies. In this example, there is a larger impetus to consider hard bargaining on the IP or its exchange for royalties. If successful – and there may only be a minute chance of that – the potential returns could be significant. It may be possible in this instance to rationalize the costs of the legalities associated with protecting the IP – including litigation, if that IP is breached.

It also needs to be considered that there may be better alternatives for the university than simply owning IP or even collecting a royalty income stream. For example, it may be possible for an agreement to be put in place such that, instead of providing the university with royalties in exchange for IP, a partner organization agrees to fund additional research or scholarships instead at the end of the research program. In many instances, the value of scholarships (which are essentially money-in-the-hand for the university) may

10.10 Intellectual Property (IP)

be a larger and more secure form of income than risky royalties down a long and convoluted track.

Another consideration in the area of IP is the issue of protecting it when breaches arise. In most countries, a breach of IP is a commercial, civil issue, not necessarily a criminal one, and therefore any patents or agreements only have force if an organization is prepared to litigate to protect them. Without litigation there is no real protection. This leads to the need to consider the cost of litigation against the cost of potential revenue. Consider Example 10.16.

Example 10.16
A university engages in a research collaboration with a small, external commercial organization. The resulting IP is patented by the university and the partner. The partner organization decides to use the IP for business purposes without paying royalties to the university, and in breach of the patent. What should the university do?

This is where the concept of IP becomes very muddy. In addition to the costs of litigation on the part of the university, there are other costs to be considered. To begin with, there is the cost to the university's reputation of becoming involved in litigation. An internationally renowned university may have a *brand value* measured in tens or hundreds of millions of dollars. That *brand value* often attracts tens of thousands of fee paying students to study at that institution, and benefactors to donate money to it. An unseemly litigation, which plays out in the media, could cause damage to that brand – well in excess of any IP value.

The other follow-on effect of litigation is the public perception of the university in society. If, as in Example 10.16, the entity that has breached the patent is a small organization and a university, being a multi-billion dollar entity, is seen to be bullying that small organization, then there is further potential damage to the university brand. At the other extreme, if the partner is, say, a large multinational pharmaceutical company, then the university has to consider whether it even has the resources to challenge that organization in the event of a breach of the IP agreement.

So, in considering these possible scenarios, it is reasonable, in negotiating any IP arrangements with a commercial partner, to consider the following as a first step:

> *"Is the university genuinely prepared to litigate to protect its IP interests?"*

If the answer to this question is no (and it often will be), then there is little value in spending inordinately large amounts of time and resources in negotiating a complex IP arrangement that will never be enforced.

Finally, in the context of IP, there is the need to look at the impact of any IP agreements upon the research student. In entering into a collaborative postgraduate research project, a student may be required to become a signatory to the IP agreement. This may involve the student signing away his/her rights to ownership of the IP, in exchange for the scholarship which has been provided for the conduct of the research. In Section 10.6.7 it was noted that research students should be counseled to take their own, independent, legal advice prior to signing any IP agreement in order to participate in a project. To do less than this is to do an injustice to the student and potentially place unnecessary restrictions on their future career or money-earning capacity.

10.11 Student Welfare

Health and safety are the two most basic aspects of research student welfare. However, there is far more to student welfare than simply health and safety.

In order to perform at maximum potential a research student needs to be:

- Comfortable with his/her surroundings/environment.
- Working harmoniously with colleagues.
- Stimulated and challenged by the research work.
- Comfortable that the work being undertaken is worthwhile and may contribute to a long-term career.
- Confident that he/she has the ability and the support to achieve good outcomes.

A good supervisor needs to track all of these factors, and they are especially important when a research student is working outside the normal university environment and in the premises of a partner organization. From the supervisor's perspective, it is in the process of examining these aspects of the research program that it becomes evident that supervision is not simply about providing technical advice on the research – it is about enabling/facilitating the capabilities of the research student.

10.11 Student Welfare

It is often the case that managers naively see their role as telling people what to do, rather than making it possible for people to actually do things. Good managers and good supervisors understand that their role is to work behind the scenes, so that the *doers* always have the right working environment and the tools they need to be at their most motivated and productive levels.

Unfortunately, it is also the case that supervisors can feel that addressing these sorts of issues is time consuming, and that the student has to learn to *manage and work independently*. These thoughts are, however, rationalizations more than reasons and it should be apparent that, in general, a research student does not have sufficient control over his/her environment or resources to resolve many issues independently. This problem is all the more pronounced in a collaborative program where the research student is an outsider to the partner entity.

A supervisor needs to be aware that any preemptive, preventative measures may involve significantly less time wastage than a postgraduate research program which has gone awry – especially one which is part of a collaboration between multiple partners.

Novice supervisors will not have had the experience of *mopping up* a postgraduate program that has gone wrong, and may therefore underestimate the scale of a clean-up task – particularly if there are multiple partners that need to be placated in relation to a poor collaborative outcome.

Logically then, a supervisor needs to put focus and energy into ensuring that the environment and resourcing for the research student are as good as they can be to facilitate a good academic outcome on the part of that student. Table 10.3 provides some useful suggestions on pro-active tasks that a supervisor can undertake in order to maintain the student's welfare during the course of a collaborative research program – especially where the student is located in the partner organization.

In addition to these rudimentary issues, there is always the need to keep a watchful eye on the mental health of the research student during a research collaboration. In particular, in such circumstances, it can be the case that a research student feels alone and isolated because he/she is neither part of the collaborating organization nor the normal university operating environment. A feeling of inclusiveness is unlikely to be achieved simply by holding formal meetings with the research student on a regular basis. Additional measures may also need to be taken – for example, organizing regular, informal events/functions at which the research student can interact and socialize with colleagues at the university – and asking the collaborating organization to do the same at its premises.

Table 10.3 Pro-active tasks for supervisor to ensure student welfare

Welfare Issues for Research Student	Pro-active Tasks for Supervisor
Comfortable with his/her surroundings/environment	• Ensure that the collaborative partner has provided a suitable and safe environment for the research student – including basic support (IT, network access, etc.) • Personally check the environment on a regular basis to ensure that it is suitable and that immediate remedies are sought with the partner organization in the event of shortcomings
Working harmoniously with colleagues	• Make efforts to hold discussions with the research student's colleagues at the partner organization • Organize informal functions to facilitate better interaction • Speak regularly to the student about his/her relationship with other staff at the partner organization
Stimulated and challenged by the research work	• Ask the student directly if he/she finds the work stimulating/challenging (n.b., this may not elicit an honest answer) • Check the student's progress against milestones regularly – if the student is completing tasks well ahead of schedule, additional challenges should be given to maintain interest
Comfortable that the work being undertaken is worthwhile and may contribute to a long-term career	• Ask the student directly if he/she believes the work is worthwhile and contributing to a career (n.b., this may not elicit an honest answer) • Organize for senior staff at the partner organization to talk to the research student about future career opportunities and interests
Confident that he/she has the ability and the support to achieve good outcomes	• Endeavor to read body language and determine if the research student exudes confidence, reticence, timidity or fear – attempt to determine root causes where necessary • Ask if the student requires additional resources or personal support to achieve outcomes

In summary, before entering into a research collaboration, a research supervisor needs to be aware that there will be significant, additional time implications for himself/herself as a result of the need to manage the research student's welfare during the collaboration. These are above and beyond those of the normal supervisory process and any technical inputs to the research collaboration itself.

11

Peer Review by Publication

11.1 Overview

It is unusual for a research supervisor to have a direct input into the examination process for postgraduate research programs – particularly in higher level programs such as Doctorates. Assessment of such programs is, rightly, left to the judgment of independent peers through assessment of a dissertation and, in some universities, through additional verbal defense of the work. It is also the case in most postgraduate research programs that a single examination process ultimately determines the outcome of that program and whether students:

- Pass outright.
- Need to resubmit minor or major revisions to their work.
- Fail the program outright.

For these reasons, a supervisor has an obligation to provide opportunities for a research student to have his/her research work independently assessed by peers, in stages – prior to any final examination. Such assessments may provide valuable insights into the strengths of the work, and any shortcomings that may need to be addressed before the work is presented for final examination.

Over and above the benefits of peer reviewed publication for the purposes of preliminary feedback/assessment, there is also the obvious need to publish as part of the normal, professional research process. To this end, having a research student submit his/her work for peer reviewed publication; interpret the reviewer assessments, and make the necessary amendments, is an integral element of the learning process.

In the context of preparing a thesis/dissertation for final examination, it is also worthwhile for a research student to highlight published work or, at the very least, the peer reviewers comments on papers that have been submitted and accepted. This provides a useful indicator to examiners that a professional process has been followed, and various aspects of the research have been independently validated/verified outside the confines of the research group in which the work was conducted.

From the supervisor's perspective – and the university's perspective – publication of research work also provides a measurable research output which is generally used in formulaic assessments of university performance. However, it is important to drive the need for publication through what is in the best interests of the research student, rather than considerations of what is purely in the *numerical* best interests of the supervisor and the university.

With these points in mind, in this chapter, we examine the issues of:

- When to publish.
- How often to publish.
- Where to publish.
- What to do if publication is not practical.

11.2 When to Publish

The obvious answer to the question of when to publish is that publication should only take place when the supervisor determines that sufficient research discovery has taken place, to the extent where findings may be of value to a broader community of peers.

This may appear to be a straightforward process but it is complicated by the length of time that elapses from the point at which a research paper is submitted to a peer reviewed journal to the point at which it is accepted and subsequently published online or in hard copy form.

A supervisor needs to be aware of:

- The sorts of journals that may be targeted for publication of the research student's work.
- The editorial policy of relevant journals.
- Typical turnaround times for submitted manuscripts in relevant journals.

Online journals give the impression that the total time from submission to publication has been reduced, but there is still the necessary and significant bottleneck in the process that is brought about by the human peer review process.

At this point, a supervisor needs to consider the milestones and timelines developed by the research student for his/her project and determine how practical these are with respect to publication. Specifically, a supervisor needs to be aware of whether he/she intends to hold to a tight project schedule – regardless of the need to publish – or work on the principle that a research student needs to stay and work on a project until it is completed – with publications. And, while the latter approach may appear to be more altruistic, it may not be practical in some university environments where there are constraints placed on the duration of postgraduate research programs – as well as research students who want to complete and get on with their professional careers elsewhere.

Realistically, in the context of, say, a 36 month Doctoral research program, the first few months will be expended on literature review; development of a methodology and supporting design of experiments or other instruments of discovery. It may be over a year before any results emerge at all – perhaps 18 months before significant information has emerged, and analysis performed, such that there is an impetus for publication. The preparation time for the paper itself may also be several weeks, taking into account time for consultation with colleagues or other relevant co-authors.

The publication process is sufficiently onerous that it needs to be factored into research student project management charts – especially because this helps focus the student's attention on the amount of work that needs to be performed and the relative urgency of getting the underlying basics completed.

A research student's research writing skills and English proficiency also need to be considered in relation to the time taken to prepare a paper for publication. A respected journal is unlikely to accept papers with poor writing and grammatical errors – even if they have some editorial staff to make minor improvements to accepted papers.

11.3 How Often to Publish

Supervisors can sometimes push their students to publish a large number of papers during the course of a postgraduate research program. In considering such an approach, it may be opportune to contemplate whether this is in the best interests of:

- The student.
- Knowledge.
- The supervisor and institution.

Each publication needs to represent a discrete and significant contribution to knowledge in its own right, if it is to have any real value to the broader peer community.

Supervisors also need to reflect on the motives for multiple publications. In particular, attempts to maximize the total number of publications from a single research program by cynical means, including:

- Dividing a single, discrete piece of research work into parts A, B, C, etc.
- Presenting the same work repeatedly with minor variations (e.g., a simulation published multiple times with variations to simulation input parameters).
- Publishing the same paper under different titles,

can be viewed as unethical and may also cause damage to the reputations of the authors and, in particular, the research student – who needs his/her reputation intact when presenting for final examination.

In a postgraduate research program, there may be natural and discrete elements of knowledge discovery that lend themselves to individual publications. If they do, then this presents a good opportunity for the research student to achieve multiple publications. However, if they do not, it is important that the supervisor does not attempt to artificially bolster the number of outputs arising from the research.

Finally, and referring back to Section 11.2, the preparation of each paper, the assessment of peer reviews, modifications and submissions of a final manuscript are all time consuming activities. It is therefore important that the supervisor ensures that each paper contributes strongly to the central theme of the research, and is not just a frivolous, peripheral attempt to improve research metrics for the sake of improving research metrics.

11.4 Where to Publish

11.4.1 General

A contentious issue in academia that has arisen with the advent of many thousands of international journals is where to publish research findings. For any given field, there may be numerous journals and conferences which may be suitable as a forum, so a decision needs to be made by the supervisor and the student.

In recent years, universities or national research bodies evaluating university performance have taken to the notion of ranking various journals. This means that, in any given field of endeavor, one publication forum may be ranked higher than another. A supervisor needs to be aware of any ranking schemes that are in place in his/her university, nationally, or

11.4 Where to Publish   217

internationally – and factor these into the decision-making process. Ultimately, however, the publication forum should be one that has:

- Direct and specific relevance to the field in which the research is being conducted.
- An editorial/review team which is highly regarded and has strong connections to the research to be published.
- An established track record of high quality publications in the field.
- Is regularly cited as a scholarly source of information in the field.
- Been used by highly regarded authors in the field.

An additional problem with modern day scholarly publishing is that it has become a large business, and there are numerous journals whose primary purpose is profit rather than knowledge dissemination. Some journals charge the authors a fee for publication and, in order to maximize profits, can publish large volumes of work having dubious value. However, the fact that a journal charges a fee for publication does not necessarily mean it is disreputable and, indeed, there are many highly reputable journals that do need to charge fees as part of their publication business model. The point is that a supervisor needs to be wary of each forum and evaluate it carefully in scholarly terms, before expending time and energy submitting papers to it.

In recent years, various groups have been assembled by scholars in order to make the process of journal selection more tractable. One of these is the *Think. Check. Submit.* Consortium *(Thinkchecksubmit.org, 2015)* which provides a range of useful guidelines and organizational links for authors to use.

For those researchers who are intent on publishing their research in open access journals, there is the online *Directory of Open Access Journals (Doaj.org, 2015)* which provides listings and links to many thousands of online publication forums. There are also numerous publication support groups, including,

- The *Association of Learned and Professional Society Publishers (Alpsp.org, 2015).*
- The *Scholarly Publishing and Academic Resources Coalition (Sparc.arl. org, 2015),*

which can provide additional support for authors.

Selecting an inappropriate journal for publication purposes can waste significant amounts of time and resources and ultimately lead to a rejection of the work. The research supervisor and the student therefore need to examine a possible journal in terms of whether:

- The journal has actually been cited in the research paper being submitted – as an indicator of its relevance.
- The authors who publish in that journal or work as reviewers have been cited in the paper being submitted.
- The research work which is being submitted is a natural follow-on (or contradiction to) other work published in that journal.
- The writing style and sentiments expressed fit in with the style of the journal.

A manuscript submission to a journal, which contains no citations to that specific journal, may be flagged by the editorial staff in terms of why it is being submitted at all, given that the authors haven't sourced any of their background material from that publication. Similarly, if a journal genuinely has noted, landmark researchers from a given field on their referee committee, then editorial staff may well question why those authors have not been cited. In other words, a journal may expect authors to demonstrate their interest in the field, and in the specific writings of that journal, in order to consider a manuscript submission for review.

Selection of a publication forum in many fields is therefore a non-trivial issue, and considerable thought needs to be given to the publication forum – and the manner in which a paper is written for that forum – not just in terms of writing style and subject field but also in terms of commitment to the particular school of thought embodied by that journal.

11.4.2 Online Preprint Servers

In recent years, internet technologies have enabled the emergence of online preprint servers for academic papers. Essentially, these are credible websites, operated by reputable universities or research establishments, which enable scholars to publish work online provided that it meets basic criteria relevant to the field. Once online, interested peers can provide immediate feedback on the published material. This provides a mechanism for resolving issues prior to expending time on a formal journal submission.

Some of the well established and utilized servers include:

- The *arXiv*, operated by Cornell University *(Arxiv.org, 2016)* and publishing information relating largely to physics.
- The *bioRxiv*, operated by the Cold Spring Harbor Laboratory *(Biorxiv.org, 2016)* and publishing information relating to biology.

These sorts of servers have become valuable tools in many fields. Indeed, in physics many new discoveries are first announced on the *arXiv*, and open, online peer review starts immediately. The benefits of such a system include

- The speed of knowledge *being made public* in the true sense of the word *publication*, thereby allaying "scooping" fears.
- The possibility to improve work quickly with rapid, extensive feedback.

The potential downside of online preprint servers is the accusation that a paper isn't peer reviewed in the traditional sense, even though in practice it may actually receive dozens or even hundreds of scholarly reviews - generally far more than the number of formal reviewers of a traditional paper.

It is evident, however, that preprint servers already play an important role in publication and this will inevitably become more prevalent. Authors should treat them exactly as they would a traditional paper submission, yet expect more thorough and more numerous reviews if the material they have published is controversial or a hot area of research discussion.

11.5 What to Do If Publication Is Not Practical

One of the most important aspects of research, as conducted in a university, is that the results should be published and made available to the broader peer community. Each piece of new knowledge can therefore be made available globally in order to potentially contribute to significant outcomes for society. However, it is not always practical to publish the findings of postgraduate research programs in conventional scholarly journals. There may be a number of reasons for this, including:

(i) The significant results from the research emerge too late in relation to the time frames for refereeing and publishing in regular journals.
(ii) The research is part of a research collaboration, and the collaboration agreement states that the results are not to be published for commercial reasons.
(iii) The intellectual property (IP) for the research has been signed away as part of some commercial agreement, in exchange for royalties or other consideration.
(iv) The research is to be commercialized by the student and/or the supervisor, and neither wishes it to become public until the IP is protected.

If time limitations (i) are the only issue in relation to publication and peer review of the work, then one obvious solution is to look at putting work on a relevant preprint server and taking note of the feedback/reviews that emerge.

From an examination process point of view, this may not receive the same level of kudos as a formal peer-reviewed publication but, nevertheless it does provide some independent validation/repudiation of the work.

The larger problem relates to scenarios (ii)–(iv) and how to create an alternative mechanism where the research student can have the various stages of his/her work reviewed by peers prior to final assessment. Clearly, no form of online or traditional publication is possible in such circumstances because it may breach contractual obligations associated with the research project.

One straightforward mechanism is that the supervisor should have the research student prepare a paper (or papers) in exactly the same way as if it was to be published in a conventional journal. For the purposes of instilling discipline into the process, a good approach might be for the research student to write the paper/s according to the style guidelines of a specifically relevant journal in the field. The supervisor can then call upon independent colleagues in the field and have them review the paper/s in the same manner as they would a formal journal submission. The qualifier to this process is that those tasked with reviewing the paper/s need to sign confidentiality agreements in line with any constraints on the supervisor/student. Additionally, if there are any binding agreements in place in relation to disclosure of research, then formal dispensation needs to be sought prior to pursuing this approach.

12

Preparation of a Thesis

12.1 Overview

The research thesis is arguably the most important document that a research student will produce during his/her candidature. In many institutions, the thesis will be the sole mechanism by which the postgraduate research program is assessed. In other institutions, where assessments include both a thesis and a defense, the thesis will constitute a critical, major component of final assessment.

The key point for the supervisor to keep in mind, in relation to the preparation of a thesis, is that it is a structured document prepared by a human for the purposes of being read by other humans. In other words, regardless of whether the thesis is in the humanities or sciences, there are large elements of subjectivity in the process – on both sides. What is seen as an exemplary work by one person may be viewed as a flawed work by another. A major task for the supervisor is therefore to ensure that, at the very least, the subjectivity in the overall process is not on the part of the student. This is a matter of rigor and discipline that needs to be instilled in the research student from the outset of the program.

There are also matters of general (not necessarily universal) convention in relation to theses, and the manner in which they are written by students. Specifically, the writing style needs to suggest that the research student:

- Is a humble learner presenting his/her work to learned, scholarly peers.
- Has learned from other scholars (literature review) and has developed his/her hypotheses in a systematic way, based upon an understanding of the existing state of scholarly knowledge.
- Has undertaken the research with an open mind, and in full knowledge of the fact that his/her hypothesis may ultimately prove to be unfounded or not substantiated.

- Has been able to develop a methodology which an independent observer would deem to be a fair, balanced and systematic means of determining the value of his/her hypothesis.
- Has been able to develop instruments, or a program of experimentation, which can be used to assess the presented hypothesis.
- Has been able to assemble/assimilate all relevant pieces of information and evaluate them in an impartial manner.
- Is an impartial observer and an unbiased writer who is prepared to acknowledge strengths in the works of others and identify and highlight shortcomings in his/her own work.
- Is prepared to provide comparisons/juxtapositions by presenting the work of others in the best possible interpretation, and his/her own work in the least favorable interpretation – in other words eliminate, as much as possible, any personal bias or subjectivity in comparisons.
- Understands the relative value/contribution of his/her own work in relation to the field of endeavor, the broader field of study, and the historical context in which the postgraduate research was conducted.
- Is able to recommend further research which may go towards addressing shortcomings in his/her own work and further extend the field of knowledge.

All of these elements of writing style need to be learned or absorbed – they are not necessarily a natural talent, and rarely are these elements learned in undergraduate programs to a level suitable for a high level postgraduate thesis.

12.2 Understanding the Thesis Readers/Examiners

Research supervisors, by virtue of their own postgraduate experience, should already be aware that a thesis is a unique document. Unlike a book, it is not aimed at a general audience. Unlike a research paper, it is not even aimed at a general readership of scholars in the field. A thesis is targeted towards a small subset of field-specific scholars who will ultimately act as examiners for the postgraduate research program. That subset of scholars may have particularly strong views about approaches that need to be taken and the manner in which analyses need to be performed. While it may be noble to attempt to change such minds, the reality is that this is unlikely – especially if those holding particular views are renowned experts in the field.

In order to write a thesis, therefore, a research student needs to understand who his/her examiners will be – not necessarily by name but by inclination.

12.2 Understanding the Thesis Readers/Examiners

If the research student has done his/her job correctly, and has conducted a rigorous literature review, then the following issues should have been resolved:

- How many schools of thought are there in the chosen research area?
- What are the strengths and weaknesses of each school of thought?
- Which schools of thought have the greatest scholarly following (citations, etc.)?
- Who are the prolific/seminal authors for each school of thought?
- Is there common ground between any of the schools of thought?
- Why do the different schools of thought exist, and what are the unresolved issues between them?
- What are the specific reasons for the research student having selected one school of thought over others?
- Does the research student's selection of a particular school of thought risk biasing research outcomes or leaving important issues from another school of thought unaddressed?

The answers to these questions should inform the research student as to who the examiners will be, and what sorts of things will interest them. Importantly, the research student needs to:

- Respect all schools of thought in the field.
- Acknowledge that all schools of thought have strengths and weaknesses – and ensure that the selection of a particular pathway for the research program has not been to the denigration of other approaches.
- Acknowledge that the pursuit of one school of thought over others is naturally limiting the scope of the research work.
- Juxtapose the least favorable interpretation of his/her research against the best possible interpretation of work in other schools of thought to demonstrate impartiality.

In addressing a particular audience, the student's objective is not necessarily to change the views of those wedded to a particular school of thought but, rather, to get them to respect the student's selection, and understand that a choice has been made in a systematic manner, and in cognizance of its strengths and limitations.

It is also important, in creating a thesis, that the research student does not seek to create a document that attempts to be all things to all people. It isn't possible to please every reader with the research choices that are made during the course of the research program, but it is important to understand which people may not be pleased with them and why.

Ultimately, however, no matter how well a research student crafts his/her thesis, it is possible that the choices which he/she has made will offend some readers who are wedded to other views. If the research supervisor and the research student have been thorough in their investigations, they should be aware of specific academics with strongly held views who, if selected to examine a dissertation, could provide an unfair assessment. Universities may, depending upon their internal processes, provide mechanisms by which research students and/or their supervisors can formally request that particular academics are specifically excluded from the examination process in order to ensure fairness.

In summary, therefore, it is important that the research student and supervisor take particular care in the conduct of the literature review in order to understand the target audience of the final research thesis. The literature review will create a picture of the target audience for whom the thesis will be created.

12.3 Thesis Structure – Developing a Thesis Template

It should be apparent from the discussions in Section 12.1 that, regardless of the discipline or specific field of research, all theses need to have some basic, underlying elements covered. There are any number of permutations in which these elements can be documented and, for each research student and project, a supervisor needs to have in mind the sort of sequence in which ideas need to be presented.

For each field of research, and its sub-disciplines, there may also be conventions that will guide the research student and supervisor towards the final thesis structure. In addition, each university will have broad enveloping guidelines and formats to which theses will need to conform.

In the absence of any other, more specific, guidelines, Table 12.1 provides a basic seven-chapter thesis template. This can either be used as a point of reference or a point of departure for the supervisor and student. The structure allows a research student to present his/her ideas in a logical sequence that begins with the basics:

- What was the objective of the research?
- When was the research conducted?
- Where was the research conducted?

and subsequently moves the reader through the details of the program, ultimately providing the outcomes:

- What were the conclusions of the research?
- What were the implications of the conclusions?
- What were the limitations of the research?
- What impact did the limitations of the research have on the credibility of the conclusions?
- What further work can be performed in order to mitigate the shortcomings of the research or extend the work documented thus far?

Table 12.1 Basic seven chapter thesis template

Chapter or Section Title	Purpose
Abstract	A short piece of text that summarizes the research program and its findings. The abstract is used by others for library cataloguing and literature search purposes
1 Introduction	A chapter designed to overview the purpose and background of the thesis, together with the proposed methodology and testing techniques. The introductory chapter needs to summarize what existed prior to the research; the specific contributions of the research and what existed after the research was completed. The chapter should also provide an explanation of how a defense of the research is presented within the remaining thesis structure
2 Literature Review	This chapter summarizes the mechanisms by which the research student identified key researchers and the major forums for publication of work. The research student needs to demonstrate, in this chapter, how he/she developed a research methodology and experimentation scheme based upon the work of peers.
3 Methodology	The methodology chapter details the proposed ideas and concept that form the basis of the investigation (the hypothesis)
4 Experimental/ Instrument Design	This chapter is critical to the research student because it demonstrates how he/she was able to develop unbiased, systematic experiments or instruments for testing the validity of the proposed hypothesis
5 Results	The results chapter provides a forum for the research student to systematically present and summarize the data arising from the experiments/studies that were performed or instruments used.
6 Broad Context Discussion	It is particularly important for the research student to take the experimental results and provide a discussion of their broader context – how they compare with other researchers and published work; how significant the results are to society, industry or a broader field of study.

(Continued)

Table 12.1 Continued

Chapter or Section Title	Purpose
7 Conclusions and Recommendations	The final chapter which summarizes, in an unbiased manner, the findings of the research, relative to the stated objectives in the first chapter. The concluding chapter should also highlight the deficiencies of the research and how these could be remedied through further investigation.
Appendices	The appendices are used as an area for storing information which is important to the arguments raised in the thesis but, because of its length, detail or complexity, would otherwise interrupt the flow of arguments in the thesis.
Bibliography/ References	A detailed listing of the sources from which knowledge and specific information was acquired.

12.4 Flow of Argument Complexity

A good research thesis ultimately tells a story about a program of research – starting with how, when and where it was conducted, and what its objectives were. Without these basic ingredients as a starting point for the document, the reader may be left confused as to the purpose and context of a long and generally complex document, which needs to be carefully interpreted in order to understand the value of the research.

A thesis reader is assumed to be a scholar with expertise in the subject field so, perhaps unsurprisingly, a common flaw in postgraduate theses is that research students try too hard to impress the reader with the complexity of their arguments. In so doing, the research story gets lost, readers get annoyed, and what may otherwise be good research gets questioned by examiners. A good technical writer appreciates that it is very difficult to get ideas across, and so a reader needs to be eased into the process.

It is also important to understand the context in which a thesis is to be read before commencing writing. In particular, a thesis examiner may agree to assess the work months before receiving a copy of it from the university. By the time the thesis examiner has received the work, he/she may have completely forgotten the background and context to the research. In authoring a thesis, the research student's task is to put the reader into the picture right at the beginning, so that the examiner can quickly come to terms with the research story that is about to be told.

The research story cannot be told by simply concatenating a series of complex technical ideas and sentences into a lengthy document. The thesis needs to contain a spectrum of information, including:

12.4 Flow of Argument Complexity

- Ideas that can be understood by a lay-person.
- Ideas that can be understood by a generalist professional in the field.
- Technically complex concepts that can only be understood by a field expert.
- Conclusions that can be understood by lay-professionals.

It is this change in complexity in the flow of arguments that makes the thesis more readable and the research story all the more compelling.

Figure 12.1 shows the flow of writing complexity from the beginning of the thesis through to the conclusions – specifically, targeting the lay-person, moving forward to the general professional, then the field-specific professional, and then the lay-professional. The same basic principle needs to be applied within each chapter of the thesis and, ultimately, within each section in each chapter of the thesis.

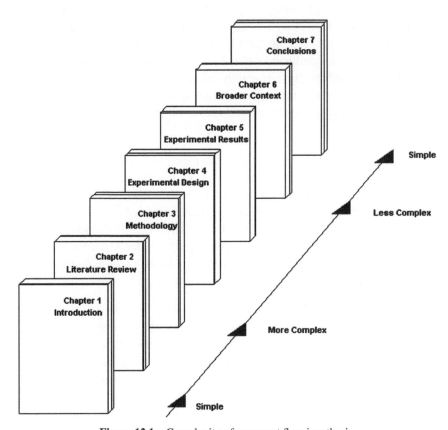

Figure 12.1 Complexity of argument flow in a thesis.

A thesis introduction needs to crafted in such a way that it sets the scene for the remaining research story. Specifically, it needs to begin with:

- The central research theme – that is, in simple terms, the hypothesis to be tested or the question to be answered.
- A statement of where the research was conducted (i.e., the university, faculty, institute, center, department or research group).
- A statement of when the research was conducted (i.e., starting and finish years).
- A statement of why the research was conducted and in what context (e.g., the research was conducted as part of a larger research collaboration).

These things are not just minor details, they are important pieces of information that the reader needs to understand in order to interpret and assess the research work and the thesis. It should also be kept in mind that, as a historical document which will be catalogued in a library and placed online, the context of the thesis is especially important. Consider, for example, that someone reading a thesis on semiconductor physics will interpret the document differently according to when the research was conducted – that is, research conducted in the 1960s would be interpreted differently to research conducted in 2016. So, the where, when and why of the research are important starting points. Consider Example 12.1 in terms of the opening sentences for a research thesis.

Example 12.1
"The purpose of this thesis is to document a Doctoral research program which was undertaken in the Jones Research Group of the Faculty of Business at the University of Northside. The research was undertaken between the years of 2015 and 2018 as part of a broader research collaboration between the Faculty and Setco Investment Partners. The research was supervised by Professor Ray Jones of the Faculty of Business. Broadly, the objective of the research was to investigate the relationship between Liquidity and Market Capitalization in medium sized enterprises."

In Example 12.1, the reader gets a simple overview of the basics of the research program, condensed into a few sentences. From that point, the reader can determine the context of everything that follows. When the thesis moves past the examination phase, and becomes a historical document, any subsequent reader can determine the relevance of the document to his/her requirements.

The complexity of the thesis logically has to increase as the document progresses because the basic purpose of the dissertation is to discuss work in cutting-edge fields. The further that one delves into the document, the more proficient one needs to be in the field in order to understand the arguments or mathematics or experimentation or modeling that are documented therein.

Moving towards the end of the thesis, there is a need for the writing style to again become more generalized – that is, aimed at a broader audience. The ability of a research student to take complex procedures, models and/or experimental results, and to present them in a concise English format, helps demonstrate a mastery of the subject matter. Conversely, when the concluding chapters of a thesis are filled with complex technical information, the reader is left to wonder whether the research student has really been able to bring together his/her findings – or, for that matter, even understands their broader significance.

A good way to check that opening and closing portions of a thesis are hitting their mark is to have a lay-person read them and see if they are able to understand them. Similarly, the overall introductory chapter, literature review and concluding chapter/s should be clearly intelligible to a lay-professional in the broader field of study.

Research students are generally very good at documenting the technically complex (field-specific) portions of the thesis – because that is where they invest the bulk of their research time – but less so at the broader concepts that need to be covered in the introduction and conclusions. These require the student to step back from his/her work and look at the broader perspectives as an outsider. The research supervisor has a role to play in ensuring that the student can do this – not purely from the perspective of the thesis but also because this is a skill required for presenting the research work to a broader audience.

12.5 The Central Research Theme

Albert Einstein is often quoted as having said that,

> *"If you can't explain it simply, you don't understand it well enough."*

Whether or not Einstein actually said it, the quote is particularly appropriate in the context of preparing a thesis, and explaining to the reader the objective of the work.

In Section 2.5.2, the central research theme was raised in the context of research outcomes, and herein this concept is examined a little further because of its profound significance in relation to the development of a thesis.

It has already been noted that a common flaw in postgraduate research theses is that research students simply don't take the time to divorce themselves from the minutiae of the research and understand what it is that they are trying to achieve in a broader context. This shows up, time and again, in research theses where the central research theme is muddled and unclear. Often, the research student sees the central theme of the research introspectively, in terms of highly complex ideas and theories, but is unable to articulate what those ideas and theories mean to the outside world. The supervisor has an important role in assisting the research student to come to terms with the central theme of the research.

By way of background, one needs to understand that it is very difficult to convey a broad range of ideas in a long document, such as a thesis, particularly when the author lacks writing experience, as in the case of a postgraduate student. Thesis writers can lose direction, and meander from one unconnected thought bubble to another, with no sense of cohesion or purpose. Irrelevant material is sometimes included for padding, while critically important information is omitted.

The key to creating a cohesive thesis is to ensure that the student author is able to encapsulate the central theme of the research within a few simple sentences – devoid of technical jargon, acronyms, mathematics or chemistry. Only after the central research theme has been enunciated – simply, clearly and succinctly, can the remainder of the thesis be created.

The central research theme is also the pivotal reference point for all the content in the thesis. Specifically, each

- Sentence
- Diagram
- Literature reference inclusion
- Set of analyses
- Discussion
- Set of results,

needs to be parsed in terms of its relevance to the central research theme. If any content is irrelevant to the theme, then it probably doesn't have a place in the thesis.

The bottom line is that there needs to be clarity of purpose – without this, a student can end up under-emphasizing the important aspects, and over-emphasizing the irrelevant ones.

In Section 2.5.2, two basic Examples (2.1 and 2.2) were provided to demonstrate the difference between an ill-considered central theme, replete

12.5 The Central Research Theme

with technicalities and jargon, and a well-considered theme which takes all the intrinsic complexities of the research and condenses them into a form which can be readily understood by a lay-person. A well-considered theme provides:

- The backbone for the research story that will unfold throughout the thesis.
- The central reference point for what the research student should include or exclude from the thesis.
- A simple, easy-to-remember frame-of-reference for the thesis reader, by which he/she can parse the validity of each claim, inclusion, exclusion or conclusion in the work – particularly for examination purposes.

The objective is to reassure the reader that the author knows what he/she is doing, because every statement in the thesis ties in to the central theme – there are no spurious inclusions and no obvious omissions. If there is a failure on the part of the student to reassure the reader – or to bamboozle them with over-use of technicalities that make thesis assessment difficult, then clearly there is a potential for a poor outcome.

Examples 2.1 and 2.2 provided evidence of the ability to take a social theme and express it in simple English grammar. However, a common concern is how to tackle a central research theme in areas of engineering, medicine or science, where technicalities are a fact of life. In this context, consider Examples 12.2 and 12.3, which look at the problem of developing a central research theme for theses which are in scientific areas.

Example 12.2 – An Ill Considered Scientific Research Theme

The photonic emissions, emanating from intrinsic semiconductors doped with Group V impurities, were modeled using Schrödinger's equations in order to verify that

$$\overline{E} = \frac{\sqrt{\varepsilon^2 - \lambda/\pi^3}}{\Phi^{j\omega\alpha} \pm \eta^{N/2}} ...$$

Example 12.3 – An Enlightened Scientific Research Theme

"The purpose of this research was to investigate the relationship between an applied stimulus energy and the emission of light, in a range of different materials..."

In Example 12.2, the central research theme is mired in a mathematical formula, rather than having been considered in terms of telling a research story in which such formulae will ultimately be revealed anyway. In Example 12.3, the author has considered the central research theme more thoughtfully, and has begun the process of explaining the research story to the reader.

In considering these discussions, as well as those in Section 2.5.2, it needs to be noted that the objective of having a simply stated central research theme is not to underestimate the intellectual capacity of the reader but, rather, to demonstrate the intellectual mastery that the author has over the subject matter. In other words, to be able to extract the simple from the complex.

Once a research student has produced a meaningful central research theme, a supervisor should encourage the student to print it out and keep the printed version visible at all times near his/her workstation – so that each activity, during the course of the research program, as well as the preparation of the thesis, is referenced against it.

12.6 Thesis Preparation Timeline

For many research students, regardless of what advice is given them prior to, and during, the course of the program, the preparation of a thesis will become a mere afterthought, following the research conduct. In these circumstances, the thesis can become a rushed attempt at a complex document – and the results of the retrospective haste will be apparent for all to see and read.

It is a challenging task even for experienced research writers to prepare a lengthy document – potentially hundreds of pages – in a short timeframe. For a novice research writer, such as a postgraduate student, it is a particularly onerous task because it needs an acquired discipline in writing style, as well as a capacity to convey ideas systematically over a complex document.

In Section 2.5 of this book, it was noted that the challenge for the research supervisor is to ensure that the thesis writing process starts early in the research program and, equally, the supervisor's editing needs to also start early. An iterative process needs to be put in place to ensure that, as soon as practical, the research student understands and achieves the required level of disciplined writing proficiency that will need to be deployed throughout the research program, and in the development of any publications relating to the research.

It was also noted in Section 2.5 that research students commonly present thesis portions to their supervisors in *draft* form, where the word *draft* is generally applied as a euphemism for undisciplined and/or unfinished work. It is not practical for a supervisor to assess a research student's capacity to write in a professionally disciplined manner unless the research student is prepared to provide a genuine – and complete – example of his/her work – not a *draft* but a legitimate attempt at a finished product.

12.6 Thesis Preparation Timeline 233

In reviewing Section 12.1 and 12.3, it should become apparent that although there are vast differences in the scope of research in varying fields and disciplines, in the context of a research thesis, the basic ingredients are common across the board. Every thesis requires:

- An introduction.
- A literature review.
- Some explanation of the hypothesis and methodology.
- A detailing of the implementation of the methodology.
- A presentation of some results.
- Some conclusions.

In fact, once a detailed literature review has been completed, and a hypothesis and research methodology formulated, a significant portion of the thesis can be written – prior to the implementation of that methodology. The challenge is in getting research students to understand this, particularly when they are eager to start on the body of the research itself. This eagerness, however, is merely pushing to one side an integral part of the research learning process – disciplined writing.

The writing style that students develop during the course of their undergraduate studies is generally not rigorous enough for a postgraduate research program. A quantum increase in quality is required. The longer the supervisor allows time to elapse before tackling the issue of creating a disciplined writing style, the more difficult the task of correction will become. Better then to tackle the problem from the outset by ensuring that the student comes up to measure before problems compound and become insurmountable.

In Section 2.5, it was noted that a logical approach to ensuring that writing rigor and discipline are tackled early is to set the research student the task of completing the introductory or literature review chapter of the thesis as soon as the initial literature review has been finished. This should set in motion an iterative process, where the student submits and resubmits the relevant chapter to the supervisor until the supervisor is satisfied that the research writing quality is at an acceptable level. The completed introductory chapter then serves as the minimum benchmark writing standard for the research student for the remainder of the research program. Needless to say, this requires a significant investment in time on the part of the supervisor – but, it also needs to be reiterated that part of the supervisory process is having the research student learn how to write professional research documents.

Time overruns on postgraduate research programs are often blamed upon the research student, or even the complexities of the research task at hand.

More commonly, however, they arise as a result of the research supervisor's failure to invest time in a disciplined attack on the preparation of the research thesis – from the outset of the program. A good supervisor should help a research student to understand that the thesis is not a minor adjunct to the research process but, rather, an integral part of it. The only way this can be communicated effectively is if the supervisor maintains a disciplined approach to submission of the various thesis elements.

A research supervisor should have, in his/her mind, a general thesis template that is particularly relevant to the field of research being investigated. A good approach is to create a thesis writing time consideration chart that can be broadly applied to postgraduate research programs under his/her charge. Table 12.2 provides an example. A more specific timeline can be developed from this by the student as part of his/her project management chart.

Table 12.2 Thesis preparation time considerations example

Thesis Element	Time-Based Action	Comments
Central Research Theme	Research student needs to enunciate, in simple terms, his/her research theme as soon as sufficient literature has been reviewed to set directions	Student and supervisor need to agree on a central research theme before the research student starts thesis writing
Introduction	Student to submit complete chapter for review 2 weeks after literature review completion	Supervisor to put in place an iterative review and modification process – until introductory chapter is of a sufficiently high standard to become an exemplar benchmark
Literature Review	Literature review chapter should be completed in full within a month of initial literature reading	Literature review chapter will need to be modified/updated regularly as new literature comes to hand
Methodology	Methodology chapter needs to be completed in full prior to commencement of any implementation work	Supervisor to review methodology chapter prior to approving design of experiments or instruments
Experimental/ Instrument Design	Experiments or research instruments need to be designed and documented in thesis prior to commencement of any testing work	Supervisor to review experimental or instrument design chapter prior to commencement of any implementation work

Results	Chapter of tabulated results to be completed as soon as results are in	Supervisor to review results chapter and determine whether additional or different work needs to be performed
Broad Context Discussions	Student needs to present his/her perspectives on broader contexts of research once results have been analyzed	Supervisor needs to review the student's analysis to determine if his/her interpretation of the results is a legitimate extrapolation of the uncovered information
Conclusions	The conclusions chapter should be completed within a few weeks of the broad context analysis	Supervisor needs to review the entire thesis from beginning to end to look at the conclusions in light of the entire investigation.

12.7 Writing Ability/Grammar

12.7.1 General

Sections 2.5.3 and 2.5.4 of this book examined the issue of disciplined writing and English language proficiency as it pertains to quality research writing as a postgraduate program outcome. Here, the issues are re-examined in the context of the thesis preparation process, as a matter of completeness.

The majority of postgraduate research theses will need to be written in English – either because candidature takes place in an English-speaking country/university, or because the majority of potential examiners in a particular field are English-speaking. This presents significant challenges for research students who have English as a second language – and for their research supervisors – and even more so when supervisors have English as a second language.

Supporting postgraduate research students with English as a second language also creates numerous ethical problems. Chief among these is the question of what constitutes the student's original contribution to a thesis under examination, and what contributions have been made by the supervisor or supporting editors. Each university should have its own guidelines in relation to supporting inputs to the writing process but there are a few points to consider:

- If the thesis is, for example, in the fields of chemistry, engineering, mathematics, medical/biological science or physics, and the examination is based upon the contribution to those fields, does it really matter if poor English grammar has been edited into a higher standard?

- Is it reasonable for examiners of theses in scientific fields to even expect a high standard of English grammar from candidates with English as a second language – especially if the quality of the research has been exemplary?
- Is it reasonable for examiners in scientific fields to ignore the quality of grammar and writing in theses and focus solely on the technicalities of the work?
- Students submitting theses in fields such as English literature are not expected to be experts in chemistry or physics, so is it fair to expect students in chemistry or physics to be experts in English writing and grammar?

As if these considerations were not problematic of themselves, consider also the submission of theses in the fields of arts and humanities, where it may be the case that the subjects relate to the technicalities of language or literature, but the theses themselves are poorly written.

Over and above these problems is the issue of technology. If it is reasonable for a research candidate to use a commonly-available, software-based grammar and spelling correction tool for his/her work, then how is that any different to seeking editorial support from an English language unit at a university?

There are no simple answers to any of these questions beyond the subjective judgment of the supervisor and the examiners, as interpreted through the window of relevant university guidelines. Each supervisor needs to make his/her own subjective decision on the level of editorial input to be provided to a dissertation, such that what is presented to examiners is a reasonable representation of the process of discovery undertaken by the research student.

It also needs to be considered that, in addition to being a tool for the purposes of student assessment, a research thesis is also a library-catalogued historical treatise in which a process of investigation is carefully documented. The thesis, as a historical document, needs to be clear, concise and unambiguous. Therefore, to allow poor grammar and spelling on one thesis, and for that document to be catalogued and placed online as a benchmark, may lead to further deterioration on the next thesis, and so on. In the worst-case scenario, this could eventually lead to documents which are unintelligible from a historical perspective. Once the process of writing discipline breaks down across the research community, it is very difficult to reinstate.

The various English language testing regimes for students with English as a second language should provide reasonable indicators of a student's capacity to write a dissertation. The reality, however, is that even for high-scoring students – and, for that matter, for people with English as a first language – the process of creating a complex, lengthy document, designed to convey elaborate concepts, is an onerous task. The English language is complex and has many subtleties – words which may be technically correct in one context may be completely unsuitable in another context – even though there is no strict, grammatical preclusion to their use.

In Section 12.6 it was noted that it is important for all postgraduate research students to commence the thesis preparation process early in the research program. It is all the more important for those with poor English skills – or those with English as a second language – to start early, because an iterative approach will definitely be required – and this will need to focus on both writing style and grammar.

It was also noted in Section 12.6 that, from a supervisory perspective, a good approach to thesis development can involve setting the student the task of completing the introductory or literature review chapter of his/her thesis as an early excercise. Keeping in mind the difficulties that a student may have with the English language, a supervisor may need to go through the first chapter multiple times in order to ensure that the thesis meets a benchmark standard that the student can use for future reference. In the case of students with English as a second language, the successful completion of a first chapter helps them to get into the natural English rhythm and banter that is used in documenting research in a particular field.

12.7.2 English Language Support Units

Research supervisors can often complain about having responsibility for a research student's writing skills but such is the nature of modern research in an international university environment. Many universities have established English language support units and editing services to assist students in their endeavors, but supervisors need to be judicious in referring students to these services.

A research thesis is not simply an English language document. In various fields of research, the basic English language is extended with:

- Technical jargon.
- Acronyms.
- Field-specific turns-of-phrase.
- Unusual grammatical expressions.

The language of a particular research field may extend well beyond basic English, and it can be unreasonable to expect generalist editorial support staff in a university unit to be able to cope with the nuances of each particular field – especially in a university that may have hundreds or even thousands of different fields of expertise.

Consider also that a person with poor English skills may submit a badly written thesis chapter to an editorial support unit – which then has to differentiate between the portions of the document that reflect poor English, and those that reflect field-specific language nuances – and all this without field-specific knowledge.

A research supervisor needs to understand that, in referring a student to an editorial support unit, they may therefore be giving that unit an insurmountable task. Moreover, what is returned from the editorial support unit may have compromised the original technicalities of the thesis – by replacing field-specific jargon with language which is appropriate in a grammatical sense but technically incorrect.

Depending upon the specific field, a better approach – from the supervisor's perspective – is to make use of an available, experienced academic (e.g., a retired one) from the same field of research. The academic can then act as an editorial support person with field-specific knowledge.

In general, however, a research supervisor will need to reconcile himself/herself to the fact that the responsibility for converting a student with poor written communications skills into one with reasonably good research communications skill will be his/hers alone. That is the harsh reality of research supervision in the modern world and, notwithstanding any university regulations about the extent of editorial support permissible in a thesis, something which supervisors will need to consider before taking on students.

Conversely, academics also need to consider the cost to society of overlooking a potentially brilliant student simply because of poor language skills.

12.8 Documenting the Literature Review

12.8.1 General

One of the most poorly written and constructed chapters in a thesis tends to be the literature review. A common error in the creation of a literature review is for a research student to treat it as nothing more than an inconvenient hurdle, which can be leaped over by presenting a collection of seemingly relevant research papers in a concatenated sequence. Ironically, instead of forming the underpinning basis for the research, the literature review chapter in a thesis can unfortunately manifest itself as an irrelevant aside, which merely strings together a collection of seemingly-relevant quotes from authors in the same field.

12.8 Documenting the Literature Review

Consider instead the literature review as the core element of the research program, and one whose objectives are to:
- Provide an understanding of the history, breadth and depth of the field.
- Establish/document a timeline of discovery in the field and the milestone events.
- Identify key/seminal research papers and scholars in the field.
- Identify various schools of thought in the field.
- Identify strengths and limitations of each school of thought and elements of disagreement/discrepancy between them.
- Determine the existing state of knowledge in a particular field and school of thought.
- Identify and formulate a program of investigation to advance the field – based upon the existing state of knowledge and field consensus of current limitations – in other words, provide an impetus for the postgraduate research.

With these points in mind, it becomes evident that the literature review – if well written – is the single most complex chapter in the entire thesis to compile and document. Unsurprisingly, therefore, many postgraduate literature reviews are not well written. A supervisor's task to ensure that they are.

Some of the typical shortcomings of postgraduate literature reviews include the following:
- Reviews are just a long sequence of quotes and ideas from various sources, with no central theme or objective that underpins their inclusion.
- Thesis authors often include references to earlier research findings without an explanation of how these are relevant to the central theme of the research – in other words, the reader is left to make the connections between cited articles and their relevance to the current research program.
- Reviews often contain information which is not relevant to the central research theme, and act as mere padding for the dissertation.
- Reviews contain the author's (i.e., postgraduate student's) opinions (i.e., personal preferences/biases) on published research, rather than a skilful balancing of strengths and weaknesses as determined by other scholars.
- Reviews often come to an abrupt halt without an explanation of how the conduct of the literature review led to – and connects with – the enunciated program of investigation.

One of the most challenging notions for a research student to accept is the fact that the literature review chapter in a thesis needs to be constructed around the central research theme – and, paradoxically, the central research theme is often derived from a review of literature. Therefore, the literature review process needs to take place through a carefully coordinated sequence of events, specifically:

- The conduct of an initial broad review of research literature.
- The development/identification of the central theme of the research to be undertaken.
- Preparation/writing of the formal literature review thesis chapter, in the context of the central research theme.

Another common problem that has emerged with the advent of technology is that research students can become preoccupied with automated referencing systems, rather than focusing on the more important issues, such as the structure of the chapter and relevance of various literature to their review. While automated referencing systems should be a boost to productivity, they can often become an end in themselves, with students expending inordinate amounts of time automating a process which they may only use once in their lifetimes, particularly if they move to a career outside the research sector. The volume of reference inclusions can also increase frivolously with automated support systems because of their ease of use, but the relevance and inclusion of what is cited can become more and more dubious.

Another concern with these systems, and online (e.g., citation search) tools is that there tends to be a case of positive reinforcement of previously cited research only, and other important work can therefore be overlooked. This is of particular concern in the case of recently published research which currently has few if any citations but may be particularly important in the context of the student's own postgraduate research.

The research student therefore needs to be circumspect and search the entirety of the literature with as many primary references as possible, and be aware of how the biases of authors, and restrictions on numbers of citations, tends to artificially select for only a fraction of the complete literature.

To use only frequently cited literature, as thrown up in computerized systems offering ranked results, is not only lazy but an incomplete set of information on which to base a review. A seasoned examiner will likely have a better grasp of the entire literature and may expose embarrassing holes in the literature review if not careful. There is also a possibility that the student may omit a pertinent reference authored by the examiner.

As far as inclusion of work is concerned, in general, references should only be included in the literature review chapter of a thesis if they meet one or more of the following criteria, specifically:

- Provision of a historical context to the research.
- Provision of a broad-based discipline view of the research field.
- Provision of a timeline of discovery and milestones in the field.
- Descriptions of the various schools of thought on the chosen research field – and differing/contradictory views.
- Analysis of strengths and weaknesses of competing schools of thought in the chosen field.
- Current state-of-the-art knowledge in the specific field to be researched.
- Published assessments of the limitations of the current state-of-the-art knowledge.
- Recommendations from other learned scholars on future research directions.

Moreover, in writing the literature review chapter, the author (i.e., student) needs to give the reader some insight into why particular references have been included. Consider Examples 12.4 and 12.5.

Example 12.4
"The research by Jones (2009) provides a useful historical context to the research undertaking during the course of this Doctoral research..."

Example 12.5
"There were two discrete, but conflicting, schools of thought on possible solutions to the problem – one was represented by the work of Smith (2001) and the other by Venkatrasan et al. (2004)."

It should not be assumed that the reader knows – or can infer – why particular references have been included – it is the author's job to explain the purpose of an inclusion.

12.8.2 Limitations of the Literature Review

Research students will rarely have the luxury of conducting work in a field in which there exists only a handful of published literature. Even if this is the case in the narrow sub-field wherein the postgraduate research is conducted, the over-arching field will be replete with published work.

Typically, the over-arching field of investigation in which a student works has emerged over decades or even centuries. For this reason, a good literature review needs to place the postgraduate research into context, as part of a timeline of discovery, and so it is important that there is sufficient breadth to provide a meaningful insight into the potential, specific contributions of the postgraduate program.

In any highly published area of investigation, research students are unlikely to be able to provide a complete review of the entire field, and what is reviewed in the final thesis may only be a small subset of the total scholarly work that has been published. A postgraduate literature review therefore has limitations, and the research student needs to address these limitations for the reader. It is important that, in the literature review presented for the thesis, a research student is able to convey:

- The timeline of discovery of the broad field.
- The overall size/scale of the field in terms of publications.
- The method by which the subset of publications chosen for inclusion in the literature review was selected in terms of its:
    - Relevance
    - Representation of the broader field of published work
- Publication areas or schools of thought which were not examined during the review, and the reasons for their omission.

In other words, the research student needs to demonstrate that the literature which was reviewed was not only relevant but also a balanced and fair representation of the total body of published work in the field.

Finally, the research student needs to be careful in any claims made in relation to his/her postgraduate work and its novelty in the field. Clearly it is not possible to prove a negative argument, and so a research student needs to exercise caution in claims. For example, it is unreasonable to make claims, such as:

> "...prior to the commencement of this postgraduate research program, no research had been conducted in this field,"

unless all work in that field has been examined – a task which is generally not feasible.

The fact that a research student has been unable to uncover published work in a particular area does not prove that the work hasn't been published. Moreover, there may be concurrent research under way in the exact same area which has not been published because of some confidentiality arrangement, or work which has been completed and is currently in the process of being published.

When a student tries to overstate the novelty of his/her own work, it naturally invites a reader to try to disprove the assertion. If the work is truly novel then the reader doesn't need telling.

Within a thesis, a research student needs to modestly enunciate the specific contributions of his/her research relative to what has already been published. However, such claims need to be made cautiously, and with a rider that a literature review can rarely be complete.

12.8.3 The Funneling Process

A good way to conceptualize the literature review – and the literature review chapter in a thesis – is as a funneling process, from which emerges the impetus for the research at hand. Consider Figure 12.2, which illustrates the process.

If the horizontal axis of the funnel represents the volume of literature available, and the vertical axis represents the relevance (i.e., subject and time

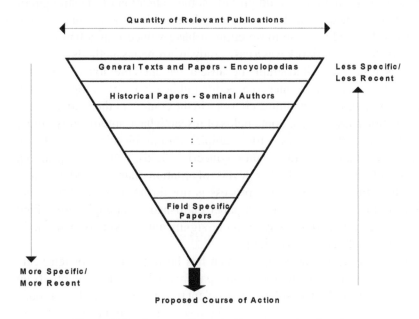

Figure 12.2 Conceptualizing the literature review funnel.

specific material) of the literature, then clearly there will be a large number of publications at the top of the funnel, and relatively few at the bottom.

Considerable skill needs to be applied in assembling the literature review because the author needs to balance the requirements for historical relevance and context, with the need to focus on material that is up-to-date and specifically relevant to the impending research. The literature review chapter in a thesis should ideally be a written manifestation of the "funnel", in order to convey the systematic investigation that has taken place.

12.8.4 Identifying Landmark Research and Seminal Authors

An important element of the literature review process is for a research student to demonstrate that he/she has been able to systematically identify landmark research in the chosen field, and the seminal authors/scholars who were responsible for the work.

Online publications and citations databases make this task relatively straightforward – notwithstanding the limitations noted in Section 12.8.1 – but it is important for research supervisors to ensure that students are not simply extracting numerical data without interpreting its significance.

Superficially, citations should provide some indication of the strength of previously published work. However, a student needs to determine whether work is highly cited because those citing it have valued it as a baseline for research, or because the work was a case-study in erroneous methods or results. Sometimes, research is cited simply because it was cited previously – and it is a convenient way for academics to expedite publication – by citing the cited.

It is also necessary to examine citations in the context of the field in which the publications have arisen. Some fields of research have significantly higher levels of citation than others – for example, various biosciences fields, on average, have higher citations than engineering fields. Generally speaking, pure research fields have higher levels of citation than applied research fields, where the next step in the process is not further research but actual implementation. So, for example, a paper published in an engineering field which receives 20 citations may be more significant in its field than one in biosciences which receives 40 citations.

It is also the case that a paper which has been published for ten years will obviously have more citations than a paper which has only recently been published – because there hasn't been sufficient time for other scholarly groups to use the work in their own research. The research student needs to acquire sufficient bibliometric data in order to make such judgment calls.

Finally, it also needs to be noted that research is not an election-based system. The *correct answer* to a field of study is not determined by the largest number of votes (i.e., citations). Each publication needs to be assessed on its own merits. There may also be a legitimate case for a student to pursue a research field in which publications have few citations – in order to get an insight into why a particular school of thought has not been broadly accepted.

12.8.5 The Literature Review as an Impetus for Research

A postgraduate research program is not intended to be based upon the pursuit of a random thought bubble, in the hope that this will lead to some miraculous breakthrough. Rather, the program is an apprenticeship in rigor – where the research student needs to demonstrate that he/she is capable of identifying research trends and their strengths and limitations – and then, based upon the scholarly work of others, develop his/her own work to extend the scope of knowledge.

A common flaw in literature reviews is that they are reviews of literature for the purpose of reviewing literature – rather than for the purpose of providing an impetus for the research. The end objective is not to demonstrate that a great deal of literature has been reviewed, but to show that the research student has been able to systematically extract a collection of ideas from other learned scholars in the field – in order to create his/her own research directions.

If a research methodology does not logically flow on from the literature review, then an examiner is likely to ask the obvious question of why the review was conducted in the first instance. There must be a logical progression from review to research method. Moreover, the research student needs to explicitly enunciate this connection between review and method in his/her thesis, rather than expect readers to interpret it for themselves.

12.8.6 Writing the Review Chapter

It is likely that a research supervisor will need to provide his/her research student with considerable support in structuring and authoring the literature review chapter – particularly in instances where students have English as a second language. There is no getting around the fact that this is a complex chapter to structure correctly, and a complex chapter to write. A useful template for a literature review chapter is provided in Table 12.3, showing the various sections that might be included within it. Supervisors can use this as a basis for developing a more field-specific version that may better suit their specific requirements.

Table 12.3 Sample template of a literature review chapter

Chapter Section	Objectives
Introduction	An overview of the end objectives of the review, the areas to be covered, and the order in which they will be covered
Review Methodology	A discussion (defense) on the method employed to ensure that the review was comprehensive, relevant and provided an impartial assessment on possible research directions
Review Limitations	A frank assessment of the strengths and limitations of the review process
Historical Perspective	A discussion on the history of the field and the timeline of discovery to highlight the context and potential significance of any outcomes relative to the total scholarly contributions to the field
Review Topic 1	Discussion of technicalities of the review area – section starts with an overview of why the topic was reviewed, and concludes with a summation of how the reviewed topic contributed towards or influenced research directions
Review Topic N	Discussion of technicalities of the review area – section starts with an overview of why the topic was reviewed, and concludes with a summation of how the reviewed topic contributed towards or influenced research directions
Summation	The summation brings together the findings of each reviewed topic in respect of their relevance to research directions
Research Directions/Impetus	A statement is made as to how the collective findings of the review specifically led to the research hypothesis and/or methodology

12.9 Balancing a Thesis

12.9.1 The Thesis Body

The primary objective of a research dissertation is to convey the story of a piece of research that was conducted by the research student, in a systematic and compelling manner. This requires judicious editing on the part of the research student (and ultimately the research supervisor) in relation to the content of the body of the thesis – and subsequently, what material needs to be relegated to appendices.

A simple rule that can be employed is that the body of the thesis is used to tell the research story, and the appendices are used to provide the detailed supporting evidence.

In balancing between material for inclusion in the thesis body and that for inclusion in appendices, consider what ultimately needs to be achieved. There are numerous impediments to telling a research story that retains the reader's interest, specifically:

12.9 Balancing a Thesis

- Long sections of mathematical formulae.
- Large slabs of computer source code.
- Large quantities of raw or untabulated data.
- Numerous graphs, charts and printouts from experiments.
- Lengthy observational descriptions of phenomena.

In each case, the author needs to ask himself/herself whether the inclusion of such material is helping the reader to understand the research story, or is it merely padding the body of the thesis to make it look thicker and more comprehensive?

The importance of each of the above materials is dependent upon the field of research – for example, a thesis in mathematics may genuinely call for lengthy sections of mathematical analysis. More generally, however, there are systematic means of dealing with these elements to ensure that they do not become a distraction to the research story that is being told in the thesis. The following guidelines may be of assistance:

- Materials which can be developed by a competent technical or professional person, and which are not unique to the research itself, but are supporting of the methodology, should be relegated to the appendices (e.g., general computer source code).
- Materials which are specific to the research and its findings should be presented in summary form in the thesis body and the details relegated to the appendices.
- Materials which clearly illustrate a particular trend or observation should be presented in the thesis body, and other similar, but not unique, items should be relegated to the appendices.

Table 12.4 provides some suggested techniques for dealing with these materials.

Left to their own devices, and in the presence of readily-available supporting technologies that facilitate the creation of large volumes of data, diagrams, images, etc., research students can be prone to *padding out* the thesis body in the mistaken belief that this will impress readers/examiners. More likely, the practice will alienate the readers because it demonstrates a lack of judicious selection on the part of the student. Research supervisors therefore need to assist and support research students with the selection of materials for inclusion in the thesis body.

A useful approach is to get the student to benchmark every inclusion in the thesis body against the central research theme. If the material in question is specifically relevant to the research theme then it can be shortlisted for inclusion, otherwise it should probably be considered for the appendices.

Table 12.4 Balancing between thesis body and appendices

Material	Method of Inclusion in Thesis
Mathematical formulae	Include only key equations in thesis body. Where derivations or proofs have been performed, include only significant steps in the body and refer to complete treatments in the appendices.
Computer source code	Only include *unique* source code in the thesis. In the body of the thesis only include small, representative or innovative sections of code and refer to more detailed coding in the appendices where required.
Raw/untabulated data	Only include tabulated summary data in the thesis body – and data which shows a trend or demonstrates an anomaly or other phenomenon – more detailed data should be included in the appendices. Raw data should only be included in the thesis if it is small in volume or highlights a specific anomaly or trend.
Experimental graphs, charts other graphical printouts	Only include graphics that demonstrate a specific trend or phenomenon related to the central research theme/hypothesis. Include a representative graphics sample if necessary to demonstrate application of the methodology or unique results – the remainder of the graphics should be in the thesis appendices – and only if specifically relevant.
Lengthy observational descriptions of phenomena	Avoid lengthy verbal descriptions – where possible convert lengthy, written descriptions into tables; timeline diagrams; graphs or charts. If descriptions relate to important interviews or experimental observations relevant to the central theme of the research, include only key points in the thesis body and relegate complete descriptions to the appendices. Do not write lengthy descriptions of phenomena which are self-evident from photographs, tables, graphs or charts.

In the context of supporting a student to make judicious selections about what to include or exclude from the thesis body and appendices, it can also be useful to have the research student question his/her own work in relation to personal reading preferences. Specifically, would the research student, as a thesis reader, want to:

- Read or decipher pages of mathematical proofs in order to understand the research story?
- Read pages of lengthy verbal descriptions, where the same information might be more efficiently conveyed with diagrams, charts, timelines or tables?
- Be distracted from the research story by lengthy passages of supporting material, which is background or peripheral material to the central research theme, rather than a key finding?

Ultimately, what should or should not be included in the body of the thesis is a subjective decision, and what the research supervisor and student may agree upon as a sensible approach may not appear that way to a thesis examiner. One has to accept that one cannot expect to get it right and please all of the people all of the time. Nevertheless, by being judicious about the inclusion or exclusion of every element of supporting material in the body of the thesis, the chances of alienating an examiner, by unnecessarily confusing the research story with padding, can be minimized.

12.9.2 Thesis Appendices

Thesis appendices are another potential source of annoyance for thesis readers and examiners. Research students often view them as a place in which padding materials can be added in order to make the thesis look more impressive than it might otherwise be. This is a naive approach and needs to be addressed by the research supervisor.

In the body of the thesis, every decision about the inclusion or exclusion of material should be assessed against the central research theme of the research, and the same rule needs to apply in the appendices. The only difference being that the appendices are a more appropriate location for detailed supporting calculations, mathematical formulae, data, source code, observations, charts, graphs, etc.

It is important not to make the mistake of treating the thesis examiner as a fool – simply by attempting to maximize the amount of information presented in the appendices. A more productive approach is to consider:

- What additional information is required in order for a reader to reproduce the work contained in the body of the thesis?
- What supporting information is required for a reader to assess the links that have been made by a student between raw information and the conclusions drawn in the thesis body?
- What is the best way of presenting the information so that an informed reader – who has read the body of the thesis – can view and assess the supporting information in the most efficient manner?

In addition to these issues, there is the question of how much supporting information needs to be presented in the appendices – and where the line falls between necessary information and worthless padding?

In Section 12.9.1 it was noted that materials which can be developed by a competent technical or professional person, and which are not unique to the research itself, but are supporting of the methodology, should be relegated to the appendices (e.g., general source code) The issue is where to draw the line.

Consider, for example, that in a thesis in the field of chemistry, there would be no obvious need to include in the appendices details on how test tubes were manufactured – unless that manufacturing process was specifically relevant to particular aspects of the research program. In an engineering research program, where a piece of software was created to perform some analysis, there would be no obvious need to include source code relating to the development of the user interface – unless that user interface was of particular significance to the central theme of the research program.

There are no hard and fast rules that can be applied, save for the basic test of assessing each piece of material for relevance against the central research theme of the thesis. Beyond this, experience is particularly useful in making subjective decisions, and therefore a supervisor has the capacity to provide valuable support to the research student in his/her decision-making as it relates to balancing the thesis.

12.10 Personal Opinions, Lazy Phrasing/Numerical Phrasing

Research supervisors can have a role to play in explaining to research students the difference between a research thesis and a book. These differences are not always apparent.

Specifically, there is a need to explain that books tend to be read by people who are particularly interested in the views, opinions and preferences of the authors. Moreover, people often read books specifically to get these personal opinions or individual insights, and not necessarily just facts. Even in books that relate to technical matters, such as bioscience, chemistry, engineering, physics or psychology, an author has the grace and latitude to skew discussions in a particular way. Book readers understand that they are getting one view of the world from a book – and that view may be neither complete nor impartial.

A thesis, on the other hand, needs to be written by a student in an impartial manner, and relate to an analysis of:

- Facts.
- Data.
- Opinions of other scholars.

A thesis is not about the personal views of the research student.

12.10 Personal Opinions, Lazy Phrasing/Numerical Phrasing

In a book, readers can show tolerance to an author who states his/her view that white is black, or black is white. These may form a legitimate opinion or the author's interpretation of the world. This is not a luxury which is generally accorded in the context of a thesis, where each and every sentence needs to be parsed in terms of:

- Accuracy.
- Contestability.
- Supporting evidence.

There is always some scope for the insertion of professional judgment in a thesis – based upon supporting evidence/information/data or other scholarly opinions – but there is little or no room for unsubstantiated personal opinion.

Research students, who are recent graduates, are unlikely to have developed the strict discipline of research writing within such constraints. Many will have written minor theses and experimental reports, but few will have had them subjected to the rigor of a major thesis as it exists in the context of, say, a Doctoral research program.

This is another important reason why supervisors need to get research students writing their first thesis chapter as early as possible, so that the discipline of research writing can be instilled before misconceptions and mistakes become ingrained, and more time-consuming to correct.

Consider Table 12.5 which shows just a few of the words and phrases which are acceptable in common writing and books, but which may be questioned by an examiner in a research thesis.

In the context of Table 12.5, consider the sentence in Example 12.6, and the issues that it creates when parsed:

Table 12.5 Commonly used words/phrases which create issues in theses

Commonly Used Words/Phrases	Issues with Usage in a Thesis
All	Suggests 100%, and needs to be substantiated with supporting data
None	Suggests 0%, and needs to be substantiated with supporting data
Most	Suggests >50%, and needs to be substantiated with supporting data
Minority	Suggests <50%, and needs to be substantiated with supporting data
Majority	Suggests >50%, and needs to be substantiated with supporting data
Small	Needs to be defined numerically
Large	Needs to be defined numerically
Vast	Needs to be defined numerically

(*Continued*)

Table 12.5 Continued

Commonly Used Words/Phrases	Issues with Usage in a Thesis
Above	In referring to previous material, a specific section, subsection, table, figure, chart or equation number should be used
Below	In referring to following material, a specific section subsection, table, figure, chart or equation number should be used
It is said that	Lazy-phrasing – author needs to define who said "it" and when
In my opinion	Thesis is not about opinions, it is about evidence and analysis of other scholarly work
Worthless	Needs to be defined explicitly
Valuable	Needs to be defined explicitly
Easy/simple	Value judgment and has emotive connotations – better to replace with a words such as "straightforward"
Difficult	Value judgment with emotive connotations – may require more formal definition

Example 12.6
"Most of the research in the field proved to be valuable, although it is said that only a minority of the authors were themselves experts at the time they performed their work."

The specific issues that such a sentence creates in the context of a thesis:

- The implication is that the writer has examined 100% of the research in the field and has determined that more than 50% has a particular characteristic.
- The author has not defined *valuable* in any strict context.
- The author has not identified who has said that none of the authors were experts – nor when he/she said it.
- The implication is that less than 50% of the authors were experts in the field, and that there is hard numerical evidence to support this claim.

Needless to say, one can become too pedantic with phrasing and, taken to the extreme, this can make the writing of a thesis an intractable proposition for a research student. Nevertheless, research students need to develop the discipline of mentally parsing each sentence as they write, in order to determine if there are any shortcomings in their phrasing.

Once parsed, the sentence in Example 12.6 could be re-written, as in Example 12.7, to eliminate the anomalies:

Example 12.7
"Much of the work in the field proved to be directly applicable to this research program, although Jones et al. (2007) observed that few of the authors were themselves experts at the time they performed their research."

Notice how, in the absence of supporting numerical information, words which had a specific numerical connotation (*all* and *none*) have been replaced with non-specific words, such as *much* and *few*. The source of the observations has been formally cited to demonstrate that it is not just the thesis author's opinion.

These sorts of writing issues should be evident to experienced researchers but they are not necessarily apparent to a novice research student. The supervisor's role therefore is to rein in the student's use of common English phrases, and instill discipline into the research writing. This is not something which can occur immediately – after all, even a recent graduate will have had more than 17 years of schooling and university, during which he/she has developed a natural writing style. The objective is not to damage the core writing style but merely to tighten it up for research purposes.

It has been noted earlier in this text that a good exercise is for a research supervisor to have the research student go through the first submitted chapter of the research thesis and parse every sentence within it, based upon the following criterion:

"Is everything within the sentence independently verifiable?"

If the answer to this question is *yes*, then the sentence can potentially remain as is. If the answer to the question is *no*, then the sentence needs to be rewritten – that is, tightened up – adding supporting data or references as necessary – or replacing specific numerical words with non-specific words.

If a research supervisor begins the task of teaching disciplined writing skills from the outset of the research program, he/she will find that students tend to be quick to adopt and learn the techniques. This makes the ongoing task of thesis review more straightforward for the supervisor, and the research story told by the research student more compelling for the reader.

12.11 Creating a Cohesive Document

A thesis needs to be more than a collection of discrete and disconnected chapters if it is to create a compelling research story for the reader. A thesis is not a novel, but some of the elements of novel writing can be deployed in order to make the thesis flow smoothly, and to create a document which is greater than the sum of its constituent chapters.

Figure 12.3 shows the end of each thesis chapter linking to the beginning of the next thesis chapter in order to create a smooth and consistent flow of arguments and ideas from beginning to end.

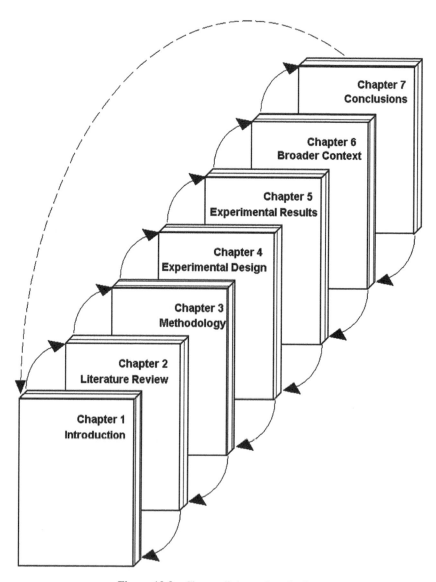

Figure 12.3 Chapter linkages in a thesis.

The question is how the linkages, shown in Figure 12.3, should occur. Consider Example 12.8, which shows the final paragraph of the Literature Review chapter, and Example 12.9, which shows the first paragraph of the ensuing Methodology chapter.

Example 12.8
"...In summary, the reviewed literature demonstrated a significant gap in testing of hypertensive medications in young adults, aged in the 20–25 group. Five of the uncovered research studies recommended that further testing be carried out. This formed the basis of this Doctoral research program, the methodology of which is discussed in Chapter 3 of this thesis."

Example 12.9
"The Literature Review chapter of this thesis documented a number of deficiencies in the testing of hypertensive medications in young adults. In particular, various research studies were cited therein which identified the need for an investigation into the effects of hypertensive drugs on young adults in the 20–25 year age group. This formed the basis of the research directions outlined herein. In this chapter, the methodology for undertaking the study is outlined in some detail..."

In these examples, the concluding paragraph of a literature review chapter is paired with the opening paragraph of the methodology chapter. One could argue that this leads to some repetition, and that the reader should be able to make such obvious linkages for himself/herself However, this does not necessarily reflect how a thesis might be read. A thesis examiner may read one thesis chapter every now and then, in between other duties, and so the thesis author needs to insert reminders and repetitions throughout the document to refocus the attention. These should summarize and bring together earlier arguments that may have been forgotten with the passage of time.

Without linkages and reminders, it is common for examiners to flag non-existent errors in the research process or thesis – simply because they have forgotten that various points have already been covered earlier in the dissertation. The worst-case scenario for a thesis that includes linkages and reminders is that examiners will argue that there is repetition in the work – a far less serious misdemeanor.

A reader should therefore not just accidentally bump into a new chapter or section, and be left wondering why he/she is reading it – and, for that matter, how or why it follows on from the previous chapter. The author's task is to ease the reader through the document by explaining how the conclusions of the current chapter lead on to an examination of ideas in the next chapter.

The summations of one thesis chapter should naturally lead into the introduction of the next chapter. The same approach should also apply to chapter sections and subsections – although, in these instances, the linking text may be as short as a single sentence.

Considering the thesis in its entirety, one of the objectives of the first chapter is to explain to the reader the structure and sequencing of the remaining chapters – so that the reader can gain some insight into how a complex story is going to unfold. This insight may also avoid having thesis examiners make unnecessary criticisms of the research process because the broad picture and process of discovery is explained from the outset.

There also needs to be a natural/logical progression/flow from chapter to chapter, and not just a staccato collection of isolated and disconnected ideas and commentary. Overall, if the research has been conducted through a systematic and logical procedure, then the thesis chapters should reflect this flow of ideas and findings.

At the end of the thesis, the concluding chapter needs to address the questions/challenges/hypotheses that were put forward in the introductory chapter and, based upon the content of the intermediary chapters, present an impartial assessment of what has been achieved.

12.12 Understanding the Broader Context of the Research

A research student, particularly in higher-level programs such as a Doctorates, needs to demonstrate to examiners a degree of maturity in terms of his/her approach to the research program. Part of this needs to be manifest in the research student's explanation of how the current research fits into the broader context of the field – both historically and technically.

The research student's ability to explain his/her work in the context of other research should demonstrate that the research student:

- Has a sound grasp of the research literature and the relative significance of his/her findings.
- Can objectively assess the specific contributions of his/her research relative to the overall history of the field.

These are not minor achievements in terms of learning the practice of research. They show that the research student can differentiate fact from fiction and personal perception – or even ego – when it comes to assessing the quality of conducted research.

A research student's enunciation of the broader context and implications of his/her work also provides a forum in which the student can ensure that the significance of his/her research is neither overstated nor understated. In other words, that the research student is able to accurately portray the facts relating to the achievements in the postgraduate research program.

It may be of value for the research student to devote an entire thesis chapter to the broader context of his/her field and the specific, relative contributions made during the conduct of the postgraduate research. From the perspective of the research supervisor, there is an important role to be played in supporting the research student to come to terms with the relative significance of the postgraduate outcomes.

12.13 Understanding the Concluding Chapter

The concluding chapter, along with the literature review, can often feature among the most poorly written chapters in a thesis. Perhaps, having devoted a lengthy period of time to the preparation of the main body of a dissertation, research students become too eager to finally bed the work down, and the conclusions suffer unnecessarily.

Common problems with the concluding chapter in a thesis include:

- Unsystematic presentation of findings/results.
- Lack of correlation between original hypothesis, results and drawn conclusions.
- Presentation of opinions for future research – rather than directions drawn directly from the limitations of the research itself.

The concluding chapter of a thesis need not be an onerous task for the student author. If the preceding chapters of a thesis have been carefully prepared, the concluding chapter should largely write itself. Essentially, the concluding chapter of a thesis should bring together the summations at the end of each of the preceding chapters, in order to present the contributions of the work – in its entirety – to the reader (examiner). This presentation should unfold in the same natural progression as has been intrinsic in the structure of the thesis.

In addition to bringing together the findings and contributions, enunciated in each preceding chapter, the concluding chapter also needs to provide some introspection on the limitations of the research that has already been presented. The limitations of the research need not be based on opinions but rather the logical enunciation of areas which have been identified during the course of the research/thesis as shortcomings.

Finally, having enunciated the findings, contributions, strengths and shortcomings of the research, the concluding chapter should contain the author's recommendations for future research – based upon those shortcomings or limitations that have been presented.

Figure 12.4 illustrates how the basic elements from the preceding chapters need to come together in order to create a systematic concluding chapter that

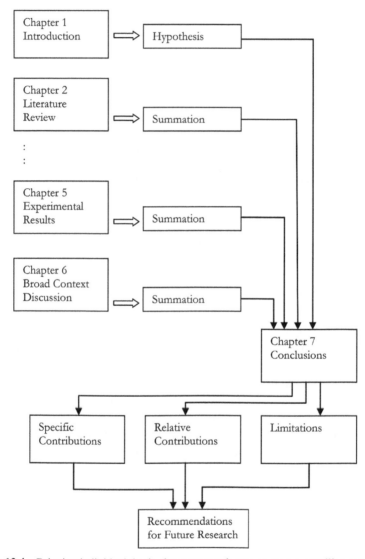

Figure 12.4 Bringing individual thesis elements together to create a compelling concluding chapter for the thesis.

presents a compelling argument, outlining the contributions, strengths and weaknesses of the research.

12.14 Understanding the Abstract

The thesis abstract has multiple functions, not the least of which is presenting a succinct summary of the work and findings that will enable:

- Library and online cataloging of the published work.
- Future readers/researchers to determine whether the thesis (i.e., research program) is relevant to their interests.
- Readers to establish the context in which the research was conducted.

With these points in mind, it is apparent that the abstract should be written, as much as possible, with the lay-person in mind, because it holds as much importance as a tool for telling some people that the research is not relevant to their needs as it does to telling others that it is. Many readers will come across the thesis abstract in an online database and will need to determine whether it is worth pursuing and reading the entire document. A well written abstract can therefore save future researchers time and energy if it is well written.

Many abstracts are replete with technical jargon, mathematical formulae, acronyms, etc. and make the task for the reader more onerous than it may otherwise need to be.

The basic elements that should be considered for inclusion in the thesis abstract are:

- When the research was conducted (i.e., commencement and conclusion years).
- Where the research was conducted (i.e., university, faculty, institute, department or center).
- Why the research was conducted (i.e., the central research theme of the thesis, in lay-terms).
- Who paid for the research to be conducted (i.e., government grant, commercial funding, etc.)
- How the research was conducted (i.e., the methodology summarized into a few succinct sentences).
- The general findings/conclusions.
- Any other anomalous/interesting features of the research that may assist a reader in determining whether the thesis is worth pursuing).

Each institution will have its own specific rules/guidelines for thesis presentation and the nature of the abstract itself. Nevertheless, summarizing a complex piece of research that has taken several years to complete – within a page or two – is often quite a difficult task, particularly for a novice writer, or for a research student who has English as a second language.

Often neglected, but especially important is the context of the research – specifically the *when, where and why* of the program. A research program which is decades old may have useful findings but they may be interpreted in a completely different way to the same type of research conducted in the present day. Also of particular importance to the context of the research is who funded the program. This is an important disclosure that goes to the core of how the research document should be read. For example, if the research documented in the thesis was related to the efficacy of a pharmaceutical product – and the research was funded by a pharmaceutical organization that was deriving a profit from that product, then it might be read differently to research which was funded by an independent government body.

The basic rules for writing the abstract are little different to those that would be applied to writing the thesis itself, and as illustrated previously in Figure 12.1. There needs to be a natural flow of argument complexity – starting with the basics (intelligible to the lay-person), then progressing to more field-specific details and ultimately returning back to basics that are intelligible to the lay-person. Consider Example 12.10:

Example 12.10
"This thesis documents a Doctoral research program undertaken at the University of Wentworth in the Faculty of Arts and Humanities. The work was undertaken at the Center for Asian Studies as part of the Bilstein Research Group, between the years of 2012 and 2016. The work was part of a broader collaboration funded by a United Nations High Commission on Refugees (HCR) research grant. The broad objective of the research was to investigate the relationship between poverty and the spread of infectious disease across three countries.
⋮
[Technical details of research program, specific findings and contributions]
⋮
In summary, the research results demonstrated that there was significant correlation between a range of poverty factors and the spread of disease in the countries that were investigated during the course of the research."

Note that by providing a preamble and summation in lay-terms, the abstract provides the broadest range of readers with the greatest opportunity for

12.15 Summarizing the Research Supervisor's Role in Thesis Preparation

determining the relevance of the thesis to their specific interests. Those who are specific to the field can read the body of the text to the required depth, and those who are not can determine whether the technical portions need to be read at all.

12.15 Summarizing the Research Supervisor's Role in Thesis Preparation

Considering all the issues raised in this chapter, it becomes apparent that the role of the research supervisor in supporting the student's preparation of the thesis is a complex one.

It is neither fair nor realistic to expect a research student – particularly one with no prior research history – to produce a lengthy document in a rigorous form without support and training. The difficulty is clearly in determining how much support and training should be given in order to ensure that what is ultimately prepared is a genuine reflection of the research student's capabilities. This is a subjective judgment that each supervisor will need to make for himself/herself in cognizance of the specific thesis structure and presentation rules of his/her institution.

The key point to consider is that the rigors of research writing need to be developed from the outset of the program in order to avoid having to untangle an unintelligible mess at the end, when the work is due to be submitted. It has been stated on several occasions in this text that this requires a real and substantial early effort on the part of the supervisor. This should put in place an iterative editing and correction process, in order to ensure that the student develops basic research writing skills to a required standard – and so that the student understands that there is a benchmark by which all future writing will be assessed. Once research students understand that there are benchmarks, below which work will be deemed unacceptable, then they will raise the standard of their writing and, hopefully, iteratively improve during the remainder of the postgraduate program.

PART III

Relevant Supervisory Issues

13

Research Misfeasance Issues

13.1 Overview

In Chapter 9, the issue of conflict resolution was covered in some detail. Specifically, in Section 9.3, the issues surrounding management of conflicts in relation to academic misfeasance were examined. Herein, we look at the issue in the context of academic implications, as well as possible remedies.

At some stage in their careers, academic staff will need to deal with academic misfeasance and its identification, reporting and ultimate resolution. This book focuses upon research supervision, and so much of what is presented here is applicable to research students and colleagues. However, research supervisors may also become aware of misfeasance that takes place at a more senior level, and all need to understand that they share a responsibility for putting an end to it as quickly as possible, wherever it occurs. A failure to expose misfeasance will only lead to more misfeasance, and possibly to an increasing extent – because it is fueled by the emboldening of those who think that they have gotten away with previous episodes.

In Chapter 9 it was also noted that the exposure of misfeasance can represent the most serious and traumatic event that an academic will experience and, at the end of it all, if an institution fails to act in correcting the wrongdoing, then an academic needs to consider the option of resignation – with all the possible career implications/damage that that will entail.

In the final analysis, universities can only fulfill their role in society if they are free from wrongdoing, be it academic, financial, personal or political. Every individual within the institution, from a first-year undergraduate student, through to the most senior university president or governing board member has an onus upon them to ensure that their institution merits the public trust that is placed upon it and its academic work. To this end, colleagues, students, the general public and other institutions will judge *deeds* in preventing and ending misfeasance as more significant than *words*.

In this chapter, we look at a range of academic misfeasance issues that might typically confront a research supervisor in the conduct of his/her duties, specifically:

- Misrepresentation of academic credentials.
- Exaggeration/overstatement of academic track record.
- Exaggeration/overstatement of research findings.
- Falsification/fabrication of results.
- Plagiarism and failure to acknowledge work.
- Theft of intellectual property (IP).
- Misappropriation/misuse of research funds.

Thankfully, these issues are not commonplace, in the sense of everyday occurrences, but they are sufficiently common that supervisors need to be aware of them.

In addition to academic issues there are also serious, personal misfeasance issues, such as harassment and bullying, which have already been covered in Chapters 3 and 9.

Each university should have its own documented specific procedures and processes for dealing with all these issues, and some of the issues may be deemed to be criminal conduct. In this chapter, therefore, the issues are examined in a generic sense, on the assumption that the details will be covered by university regulations or national/regional legislation.

It should also be apparent that the approach to be taken in managing issues of misfeasance needs to consider the nature of the alleged perpetrator. Specifically, a research supervisor has formal responsibility for the welfare of a research student, so if it is the case that a research student has engaged in misfeasance, then the supervisor needs to act with caution in order to:

- Provide natural justice.
- Support the student while any investigation is under way.

The same level of *duty-of-care* may not necessarily apply if the alleged perpetrator is more senior than the supervisor, because the supervisor has no formal responsibility for management of that individual.

In all cases, the research supervisor needs to be aware that any accusations which are made could have serious legal implications in terms of defamation – including libel and slander. It must also be well noted that defamation laws vary greatly from country to country. In some countries, the fact that an allegation is proven to be true can mitigate against defamation. In other countries, even if an allegation is proven to be true, any statements raised may still be deemed

13.1 Overview

defamatory if they do not satisfy an additional benchmark of being in the public interest.

The utmost care therefore needs to be taken to ensure absolute confidentiality with any privileged information which arises during the course of an investigation, and that discussions on such matters are strictly reserved for relevant university office-holders. It is also worthwhile to check on the specific defamation laws which apply in one's own country before pursuing any action on misfeasance.

Table 13.1 presents an approach that might be considered when dealing with misfeasance within the university.

In terms of pursuing any issues of misfeasance in a professional manner, the starting point is always *the assumption that the person involved is innocent, and that the objective is to gather sufficient information to determine why anomalies exist – and to correct them.*

It is particularly important not to use hearsay or the mere appearance of misfeasance as an excuse for a witch-hunt, or to settle scores that have arisen as a result of personal antipathies with research students or other staff members. This sort of behavior will inevitably backfire and may lead to serious charges against those making vexatious claims.

Table 13.1 Steps for dealing with misfeasance

Stage	Issues	Comments/Considerations
Awareness	• Be aware that misfeasance can occur in the university environment. • Be aware that there can be many innocent reasons for perceived misfeasance – including simple administrative errors or oversights.	• Accusations of misfeasance can have extremely serious consequences for both the person reporting the incident and the person allegedly involved. • Never act on mere hearsay, or make accusations unless sufficient objective facts and data exist to warrant investigation.
Check on university regulations and procedures	• Universities have regulations and procedures for dealing with misfeasance, including specific committees (e.g., research ethics) and specifically nominated office-holders.	• Do not attempt to take action in isolation or independently of constituted committees. Any personal involvement in the issue may need to terminate once an appropriate office-holder or committee is formally advised of misfeasance.

(Continued)

Table 13.1 Continued

Stage	Issues	Comments/Considerations
Collection of facts and data	• Data, facts, written correspondence, etc. need to be assembled before raising any claim of misfeasance.	• Do not take actions based upon rumors, hearsay or other subjective information. Evidence should be objective and irrefutable.
Confidential meeting with person involved	• If the person involved in the misfeasance is a colleague then consider a confidential meeting to discuss the issue of misfeasance.	• Do not make accusations in the meeting. Carry out discussions on the assumption of innocence and with the intention of resolving irregularities. • If necessary, present facts, data or other information that need to be addressed
Confidential meeting with university office-holders	• If the person involved in misfeasance is not a colleague, or if personal private discussions have failed, organize a meeting with an appropriate university office-holder or committee representative.	• Present only facts, data, written correspondence at the meeting – do not venture opinions on the subject. • Leave the authorized office-holders or committee members to pursue the matter, providing input only if requested.

In commencing any investigation into misfeasance, consideration also needs to be given to the long-term relationship between the person allegedly involved and the person making the claims against them. Unfounded accusations can destroy careers, future relationships and, if widespread in a research group, department or faculty, can render the organization dysfunctional.

13.2 Misrepresentation of Academic Credentials

Universities are entrusted by the public – and sometimes through formal national/regional legislation – to be the gate-keepers of academic integrity, specifically as it pertains to the issuance, validation and verification of any awards that are granted to students. The misrepresentation of academic credentials is therefore a serious issue that universities need to confront whenever and wherever it is identified – and it is not an exceedingly rare phenomenon.

There are clearly differing practical ramifications associated with the type and manner of misrepresentation. For example, the falsification of a medical, dental, veterinary science, law or engineering degree could have extremely serious and direct ramifications. The falsification of an arts degree in history

13.2 Misrepresentation of Academic Credentials

may have less obvious consequences. Nevertheless, from the perspective of an institution, all such breaches are a serious matter.

For a research supervisor, entrusted with the supervision of a postgraduate student, the most obvious responsibility is to ensure that an impending student has the qualifications that he/she claims to have, in order to undertake the advanced qualification. Good universities will already have procedures in place to screen all applicants for higher degrees by vetting their stated qualifications, and requesting original or certified transcripts from the originating institutions. If the university does not have centralized screening of higher degree candidates then, self-evidently, the responsibility for validating stated qualifications falls to the supervisor.

In any given university, a significant proportion of higher degree candidates come from the same institution, so validation of claimed qualifications is a relatively trivial matter – even more so when the supervisor is already familiar with the individuals as undergraduate students. However, the process is considerably more complex when higher degree applicants come from other institutions, and especially for those who are international students.

The practice of insisting upon original documents as a means of validating qualifications only has limited value in an environment where sophisticated software and printing technologies are widely available at negligible cost, and can produce convincing replica documents. A further consideration is that there is little value in asking for an original document from an international institution when nobody has any experience with what the genuine documents from that institution actually look like – there is no point of reference.

It is common practice in higher degree programs for an institution to request that research students present original transcripts and testamurs from their own universities in order to be given candidature. In order for a graduate to request original academic transcripts and testamurs from his/her university, it is generally necessary to produce identification – however, this is not the case in all institutions.

The situation is further complicated when graduates have a common name, such as *John Smith*. There is nothing to preclude a person with the name *John Smith* from presenting at a university where another *John Smith* has actually graduated, in order to acquire an original set of academic transcripts that can be used for misfeasance. In practice, therefore, unless photographic identification is matched with student photographic records, it is difficult to unequivocally validate the credentials of students from other institutions.

One systematic and rigorous approach is to ask the applicant to provide a reference from an academic supervisor/lecturer from the preceding institution and then, with the student's permission, contact the referee to ensure that the student is who he/she claims to be.

If one is to be pragmatic about the issue, as long as a research student is able to perform the required research tasks and publish work then, in a practical sense, it may mean little whether they have their stated original qualifications or not. However, in the context of universities acting as gatekeepers, it is important that a postgraduate researcher is a person who has the equivalent skills of someone who has already graduated. After all, the implication of a higher degree, such as a Doctorate, is that the recipient has formally graduated with the learning and knowledge implicit in an undergraduate program, and has then taken on advanced studies. The fact that someone is capable of undertaking advanced studies without having passed the prerequisite undergraduate materials is then neither here nor there.

From the supervisor's perspective, if it becomes apparent that a research student has misrepresented his/her credentials in order to get into the program, then clearly formal action needs to be taken. Most universities will have processes in place to deal with these sorts of issues, and the matter may well be taken out of the hands of the supervisor and passed through to a more centralized procedure. In the first instance, however, the supervisor has a responsibility to ensure that he/she has all the facts before raising the matter elsewhere.

It is also important for the supervisor to raise the issue tactfully with the research student prior to escalating the matter, to determine whether there are any special circumstances related to anomalies in claimed credentials – the presumption of innocence. For example, in the case of international students in some countries, first and last names are often reversed, and checks may fail to find qualifications where they do in fact exist.

There are other situations where a supervisor may identify misrepresentation of credentials – these may be from a colleague or even a more senior staff member. Regardless of the seniority of the person involved, the issue of misrepresentation of academic credentials needs to be addressed. And, if there is an appearance of misrepresentation because of some anomaly in the person's history (e.g., his/her institution changed names or has became part of another institution – both scenarios being relatively common), then this also needs to be cleared up. In an organization where the core business relates to the issuance and integrity of credentials, it is clearly critical that all staff and students have the credentials they claim to have and that any anomalies are eliminated, so that the public can have confidence in the institution.

13.3 Exaggeration/Overstatement of Academic Track Record

Finally, it needs to be reiterated that, in terms of managing anomalies with research students, colleagues or more senior staff, it is *imperative* that strict confidentiality is maintained. If it eventuates that there is an administrative error rather than misfeasance, then any public declarations of impropriety could lead to legal defamation charges of libel or slander. For these reasons, issues relating to misrepresentation should only be discussed with the person involved and with the authorized office-holder of the university, whose task it is to provide oversight of such issues.

13.3 Exaggeration/Overstatement of Academic Track Record

Few research students will have any significant research track record when they commence their postgraduate program, and so the issue of exaggeration/overstatement of track record is generally a moot point. Nevertheless, as they progress through their research program there will be increasing occasions and opportunities for them to present their curriculum vitae – not the least of which will be towards the end of the program as they seek to obtain employment.

Research students graduating from a university carry its imprimatur and so there is some onus on the supervisor to ensure that what is presented to future employers – insofar as it pertains to the postgraduate research program – is an accurate reflection of the achievements and activities that took place. Of course, supervisors do not necessarily have the right to view what graduating students are presenting to future employers – unless the supervisors are intending to act as referees for the candidate. However, in those instances where supervisors are privy to claims made by the research student, it is important that the claims are accurate.

The practical reality is that the modern curriculum vitae serves two purposes – one is in the depiction of a career and the other is as an advertorial tool to promote the individual. Some latitude needs to be given to claims in consideration of the latter function, but there is a need to ensure that a misleading picture of the research student's capabilities is not created as a consequence – and seemingly endorsed by a supervisor who may act as a referee.

It is also entirely likely that, during the course of a career, an academic will come across various instances of exaggeration, overstatement and embellishment of claims by other academics. As gatekeepers of knowledge, universities and their staff all have an innate responsibility to ensure that what is presented by the university is accurate and independently verifiable. To the extent that academics are made aware of misleading claims, arising from overstatement, there is some onus to correct the record.

Some universities have implemented online databases in which academic staff enter specific achievements (publications, citations, patents) as they arise, and other material related to staff – such as research grant data – is automatically populated by centrally maintained data records. In addition to this are included qualifications which have been independently validated by the university. Collectively, these automatically generate a *pro forma* curriculum vitae for each staff member, which has a reasonable level of accuracy. The *pro forma* structure leaves little room for embellishment and provides a useful mechanism by which outsiders can check claims made by academics. However, without such systems being in universal usage, personally prepared documents are still commonplace, and so there is a need for academics to be alert in regard to misleading overstatement.

Typical areas of overstatement in academic résumés include overstatements in relation to:

- Publications or contributions to publications.
- Citations.
- Research grant funding.
- Patents or involvement in development of technologies for patents.
- Start-up companies and involvement in start-ups.
- Participation on government, research or business/industry boards and committees.

These are all in addition to another potential source of overstatement – that which pertains to the outcomes/ramifications of research itself. This will be dealt with in Section 13.4.

The question, however, is what can be done in order to identify, prevent or minimize overstatements? This is a complex issue because – by definition – overstatements of achievements contain some elements of fact/truth, which have been stretched. For example, an academic claiming participation in a start-up company, or the development in technologies which have been patented, may well have had some involvement in these activities – however, a fair-minded, independent observer might dispute that those contributions warrant any real claim of participation.

To some extent, challenging these anomalies is a thankless task, and one which will clearly cause animosity with the person making the overstatements. There is also the possibility of litigation arising from making allegations of wrongdoing against an individual, which are difficult to substantiate with hard facts/data.

The only real tool that academics have to reduce this type of misfeasance is to ensure that, where practical and not defamatory, hard data is made available to decision-makers who might be adversely affected by the overstatement of others. Opinions and other subjective information relating to overstatement achieve little more than to inflame the situation because they are, in effect, being used to challenge other opinions and subjective information.

In the absence of verifiable numerical data, official documents, sworn statements, etc., the task of challenging overstatement is difficult in a practical sense. Overstatement is certainly not the clear-cut breach of faith that exemplified in falsified formal credentials. So, while there is an onus on all to protect the integrity of knowledge, in many cases one has to be pragmatic enough to recognize that the tools to achieve this protection may not always be available.

Calling out staff or students in relation to overstatement is fraught with risks (including defamatory slander and libel charges) because any public declarations about overstatement are difficult to defend if overtly challenged by the offended party.

13.4 Exaggeration/Overstatement of Research Findings

In many fields of endeavor, the results arising from research are clear, objective and unequivocal. Notably, these tend to be hard sciences (physics, mathematics), engineering and chemistry. In other areas of study, including business, economics, humanities, social sciences, biosciences, etc., there are elements of subjectivity, as well as the confounding problem of determining/interpreting the significance of statistical outcomes – beyond the strict statistical definitions. Over and above these basic issues, there is also the challenge of interpreting the importance of a piece of research within a historical or field context. This, combined with the tendency of universities to provide career advancement, based upon perceptions of the importance of an individual's research, leads to the problem of overstatement.

Exaggeration and overstatement can take on various forms – and some of these go beyond an interpretation of the significance of the research results themselves. Sometimes, the overstatement can spill out into the public domain through exaggerated claims of the implications of particular research outcomes.

A good example of overstatement would be one where the news media reports – based on an academic's assertion – that some public-good outcome will be achieved *within five years* – when it is already known that the actual, current state of affairs is only a piece of basic research with some positive outcomes.

In the overall scheme of research and development, basic research generally needs to be translated into applied research, where the basic research is aligned with a practical application. Subsequently, there may need to be pre-competitive research, development, certification or approval, design for manufacture, manufacture, marketing, logistics, distribution and sales. All of these could take a decade or more to complete, so there may be no chance of a practical transition from basic research to end product, *within five years*. In other words, what has emerged in the news media as a result of information provided by a university is an *inaccurate overstatement* of the reality of a situation.

This sort of overstatement can come about as a result of an enthusiastic university marketing department overstating already overstated academic claims, to convert them into media releases, and the news media again overstating these in order to give a story greater impact. The end result is that the reality and the public perception of what has occurred in the university research are completely incongruous.

Some may view this example of overstatement as relatively harmless, in the sense that it may spur greater participation in university research or greater benefaction, but the reality is that it is prevarication – a falsehood which is allowed to emanate from a university through what may initially be perceived as a small, white lie.

Stakeholders in the university system – and that includes students, staff and the public – need to have confidence in that system. If a university claims that, for example, a pharmaceutical product to cure a particular ailment will be available *within five years*, then the university and the staff member need to be held to account for that claim. While one white lie may slip through unnoticed, ultimately, if academics and universities continually overstate the significance of their research, then the institutional goodwill, which has often been built up over centuries, will be damaged – and so too will be the goodwill of other institutions who may have had no involvement in the overstatement.

All academics have a role to play in ensuring that information which emerges from a university is, to the best available knowledge, entirely accurate – or, at the very least, a genuinely held subjective opinion which may be independently upheld or supported by a body of other learned scholars.

13.4 Exaggeration/Overstatement of Research Findings

University research is generally at the cutting edge of knowledge, so there are always opportunities for outcomes to be contested as other information emerges, but there must be underlying integrity to the process – otherwise all institutions become discredited – and, once credibility is lost, then it is almost impossible to regain.

The difficulty with challenging overstatement in research findings is the same as the difficulty in challenging overstatement of personal academic achievements. That is, the exaggerations often run so close to the facts that they are rarely complete fabrications. For example, it may be possible that a current piece of basic research will lead to a pharmaceutical cure for an ailment, *within five years* – however, this may also be a highly improbable scenario. How does one challenge such exaggerated claims when there is a kernel of truth buried in them?

One possible approach is provided by tools, such as PubPeer.com *(Pubpeer.com, 2016)*, which allow other researchers a convenient mechanism to provide rapid feedback on an article they may decide lacks credibility. So, in many ways, exaggerated claims, especially in mainstream media outlets, should be expected to provoke negative consequences for the researcher/student. It also needs to be noted that tools such as PubPeer cross reference with large, professional databases such as the National Institutes of Health (NIH) PubMed system. Thus, in exchange for a minor flourish in the mind of the general public, an unethical academic may find a major professional backlash.

Another approach to minimizing the levels of research overstatement that leave the university confines is to ensure that work always receives local peer review – within the research group, center, department, institute or faculty. Peers need to ask questions and get answers before research moves into an arena where it is not reasonable to expect practitioners to have the skill to challenge anomalies (e.g., the news media).

For example, in the case of an academic claiming to have generated a piece of research for which a commercial product will emerge, *within five years,* some reasonable peer questions could include:

- If you read a similar claim by one of your close competitors, what would you think of it? If the opinion of those in your field would be negative, what consequences might result?
- Have you ever worked previously in this area of business in a professional capacity?
- Do you have commercial partners signed up?

- Do you have seed funding or venture capital?
- How long will the development phase take?
- How long will it take for certification/approval?
- Who will fund the certification/approval process and how much will it cost?

This basic questioning should be applied as a normal matter of principle within research groupings, as part of a Socratic approach, which should be an integral part of university learning. Colleagues always need to ask questions in order to encourage the proponent of an idea or claim to engage in a deeper level thought process about his/her assertions.

The Socratic Method is particularly important in the case of research students who make claims about their research – especially overstated ones. The objective is not to embarrass, accuse or humiliate the research student but merely to get the student to consider the broader implications and possible flaws in his/her thought process.

At the end of the process, insofar as it pertains to research students, research supervisors have a direct responsibility to ensure that any claims made by the student – through the auspices of the university – do not extend into the realm of overstatement or exaggeration.

13.5 Falsification/Fabrication of Results

There are many shades of gray in research, and there are always people who can take advantage of these, as is apparent in various forms of overstatement. However, there are no shades of gray when it comes to the falsification or fabrication of results. This is a clear-cut case of academic fraud, and all academics have a responsibility to ensure that the practice is called out for what it is, as soon as it is identified.

The practical problem is that falsification or fabrication of results is often difficult or, at the very least, extremely time-consuming to identify. Unless one is watching over the shoulder of another researcher when experiments or surveys are conducted – or results entered into a spreadsheet – then how can one possibly tell if results have been falsified? The answer is that one generally cannot identify falsification or fabrication unless:

- An individual who was present during a research procedure has first-hand knowledge of the incident.
- An individual has been asked by his/her supervisor/manager to actually fabricate or falsify results.

- The fabricated/falsified results appear to be outside the boundaries of what one would normally expect for a particular type of research – that is, too good to be true – data fabrication is then detected through statistical anomalies.
- People are aware that the facilities/resources required to actually undertake a program of investigation are simply not available and, therefore, the only way to get results is by falsification/fabrication.
- Rumors circulate within a research group, center, department, institute or faculty that fabrication/falsification of work has taken place.

Ironically, despite the fact that one would assume that individuals involved in falsification/fabrication of results would be intelligent enough to cover their tracks, one or all of the above factors often come into play in uncovering academic misfeasance of this kind.

The basic question of suspicion can come down to why a researcher would want to fabricate or falsify results in the first instance. A genuinely high caliber researcher may be unlikely to engage in such misfeasance, because exposure of fraud at this level would irretrievably damage a hard-earned reputation. But then, as has become apparent from numerous public *outings* of academic fraud around the world, some people have only become eminent because of that fraud. The picture is therefore more complex than might initially appear to be the case.

For those who have no real track record of success, limited abilities, and a penchant for academic career-climbing, there is perhaps a greater temptation to engage in unethical conduct. Ultimately, if wrongdoing is exposed, then no real career or eminence has been damaged because without the fabrication/falsification there probably would be no career and no eminence.

The process of peer review does not entirely ameliorate the problem of results fabrication/falsification. After all, how can a journal reviewer – perhaps in a different country – possibly determine whether a research paper is based on manufactured or falsified data? The only real checks that can be made on such research integrity are those which are conducted in the immediate surrounds of the individuals undertaking research. Therefore, localized peer review within a research group is a critical, but by no means fool-proof, counter-measure.

There is also, increasingly, a requirement to make raw data available (or to offer it if requested) as a condition of journal publication. This increases that chances of a dishonest researcher being caught out eventually.

In the long run, if fabricated research is published – and nothing comes of the published work – then the damage inflicted is limited. If, however, other research groups expend time, resources and energy pursuing a field of investigation, based upon falsified work published by others, then the implications are far more serious. Even in the best case scenario, however, an individual who fabricates/falsifies research can achieve unwarranted career gains based upon his/her misfeasance.

Needless to say, universities have in place punitive measures to deal with this form of misfeasance once it is detected but the more important issue is how to minimize its occurrence in the first instance.

It cannot be overstated that localized peer review is absolutely paramount in this regard because the further that research moves away from its immediate surroundings, the less scrutiny that can be applied. Localized peer review, however, requires:

- A critical mass of researchers within a group, center, department, institute or faculty.
- People who are willing to actively participate in a Socratic learning approach and genuinely challenge results that are presented.
- A dynamic learning/research environment where having one's ideas challenged is welcomed by all participants.

These appear to be basic requirements for any university but, in practice, in an environment where staff are all seen to be competing against one another for goals such as tenure, promotion, salary bonuses, etc., they are quite difficult to achieve. They require ongoing, sustained effort and input from all players. In the final analysis, an ability to achieve and sustain such an environment is the hallmark of a good university.

13.6 Plagiarism and Failure to Acknowledge Work

A common form of misfeasance that research supervisors need to deal with is the allegation of plagiarism – or failure to acknowledge earlier research – particularly by research students.

There are two potential causes for this misdemeanor. The first is intentional wrongdoing on the part of the individual concerned. That is, an attempt to exploit the work of others for inappropriate personal gain. The second, more common in the case of research students, is a lack of training or awareness in relation to referencing and citations. Clearly, the first cause is a serious ethical issue requiring formal investigation and punitive measures. The second cause requires reflection and consideration on the part of the research supervisor.

In looking at the second cause in more detail, one has to understand that undergraduate learning approaches vary from university to university and, more particularly, from country to country. Some developing countries rely upon rote learning even at university undergraduate level, and there may be little regard for citing work, because it is just automatically assumed that undergraduates are rote learning from the work of others anyway. However, once students enter into postgraduate research programs, there is a responsibility on the part of the supervisor to ensure that students understand the requirements and methods of appropriate referencing.

In recent times, online software checking tools have been used extensively to uncover plagiarism in the university world. These need to be used with some caution because they tend to pick up commonplace sentences and phrases, and allude to plagiarism when in reality it is more a question of people in a particular field developing, and putting into common usage, their own vernacular.

It also needs to be kept in mind that, in any given research field, there are only limited ways in which some concepts can be explained/expressed, and it may be too pedantic to expect a researcher to uncover the *originator* of any particular explanation, especially when that explanation is really just a blatant statement of the obvious to anyone practicing in that field.

Returning to the first cause – that is, a willful attempt to utilize the work of others for personal gain – the remedial pathways are similar to those that need to be adopted for any other form of misfeasance, specifically:

- Identify formal university definitions of plagiarism, and the processes to be followed in order to resolve any accusations.
- Act on the presumption of innocence, unless there is hard, unequivocal evidence to the contrary.
- If practical, meet with the offending party and offer to work with them to determine the cause of any irregularities.
- Maintain strict confidentiality at all times, and do not discuss any allegations with parties other than official university officeholders.
- If sufficient hard evidence materializes, hand over any formal investigation to authorized university officeholders.

In all these steps, there needs to be a consideration of the motive of the alleged perpetrator – in other words, is there any apparent motive for the misfeasance or is it more likely to be an issue of carelessness or oversight? In particular, in the case of research students, one needs to skew considerations towards carelessness and oversight before act of malice – in which case, the remedy may not be punitive but educational in nature.

13.7 Theft of Intellectual Property (IP)

In a strict business sense, the bulk of IP that is generated in the university environment has little or no real value – particularly if there are no instruments available (i.e., venture capital, partnering, sale of intellectual property for royalties, etc.) to convert that IP into a large financial windfall. Every now and then, a university discovery leads to a significant financial win but this is more the exception than the rule.

IP does, however, have other benefits, not the least of which is that it may be used as a tool to help an academic climb the career ladder by virtue of eminence, prestige, international renown/recognition, awards, etc. At a much deeper level, many academics are profoundly committed to their research and betterment of society or technology. The money which academics forgo in terms of direct income is offset by the personal satisfaction of peer recognition and esteem. So, IP does have value across the board, albeit not necessarily a direct monetary value.

The theft of IP therefore goes to the heart of what academics strive for in their work, and the loss that is incurred is far greater than simple monetary short-changing. Theft of IP is effectively an assault on the passion, commitment and dedication of one academic by another.

It is difficult to think of circumstances in which the theft of IP is unintentional. Generally it is a malicious and willful act, and universities should have procedures in place to deal with it. There are some mitigating exceptions, however, that need to be considered.

Consider that academics in a particular field of research generally have similar undergraduate training. They have studied the same subjects, the same theorems, and have often learned similar approaches to problem solving. When they enter into postgraduate research and, ultimately, professional research, academics:

- Read the same journals.
- Follow the same investigative pathways that have been trodden by their predecessors.
- Apply similar reasoning or logic to extrapolate pathways.
- Are surrounded by likeminded individuals working with a common mindset.

It would be surprising therefore if, every now and then, two disparate research groups – perhaps in different institutions or even within the same institution – didn't ultimately come to the same conclusions about a particular research pathway. So, some consideration needs to be given to the possibility of serendipity taking a hand in creating similar or identical research outcomes – rather than a willful act of IP theft.

Beyond the realm of the happy coincidence, and where the IP theft is clearly willful, it can manifest itself in various forms, including:

- Academics publishing the work of others as their own, without acknowledgement.
- Individuals, and particularly businesses, using technologies or approaches developed by academics and protected by various instruments (e.g., patents or trademarks) without acknowledgement or payment of royalties.
- Academics claiming the work of others as their own career achievements in order to achieve a promotion.
- Academics submitting the work of others for the purposes of achieving awards/prizes.

If an academic views such an act as willful then little can be achieved by confronting the alleged perpetrator directly, and redress needs to come through whatever formal processes are made available by the university for resolution of such disputes. If the disputes involve financial and business interests beyond the integrity of knowledge itself, or are with an external party (e.g., business enterprise), then the matter can become a civil legal proceeding. At this point, the university – should it choose to defend the property of one of its staff – will generally undertake a cost/benefit analysis in relation to defending the IP. At the very least, the institution needs to determine if the defense of the IP is critical to its reputation and ongoing research interests.

If a university elects not to pursue an infringement of IP, then an academic may need to consider whether he/she pursues the matter in a legal sense as an individual. It is unlikely that an academic, as an individual, will be able to rationalize litigation in a cost/benefit sense any more than the institution is able to. Nevertheless, there may also exist issues of honor and integrity that an academic wishes to protect by virtue of a civil litigation.

13.8 Misappropriation/Misuse of Research Funds

Research funding in universities comes from a limited range of sources, notably:

- Recurrent university income from fees and government funding.
- Income derived from university endowments and trusts.

- Project research funds obtained on a competitive basis from national funding bodies (health and science).
- Project research funding from benefactorial donations.
- Project research funding from business/industry for the purposes of undertaking a collaborative research program.
- Contract research and development funding from business/industry in order to achieve a particular applied research outcome.

In each case, the terms under which the funding can be expended by an academic staff member are enunciated in either a written, legally-binding contract or through rules, regulations, procedures or national/regional legislation. In other words, regardless of the funding source, there are either career, civil or criminal sanctions available to the entities which provide the funding – if the funding is misappropriated or misused by an academic staff member.

Typically, research grants are awarded for expenditure on a limited range of items, including:

- Procurement of research students, contract research staff (e.g., postdoctoral researchers), or technical and administrative support officers.
- Payment for time-release of the academic staff member – often to the university – so that the staff member can participate in the research.
- Payment for resources, equipment or consumable items related to the research project.
- Funding for travel and attendance at relevant conferences.
- Local incidental travel and sundries.

If the purpose of a research grant is to establish a new or larger entity, such as a research center or institute, then of course there is greater flexibility in the discretionary expenditure of funds.

Modern accounting systems readily accommodate the sorts of functions outlined above, and provide tracking of actual expenditure against specific expenditure types. Nevertheless, there are always those who are naive enough to believe that they can get around the sophisticated accounting software in order to misappropriate or misuse funds. The majority of wrongdoers in this field of misfeasance are eventually caught out by the system, through audit, and the punitive consequences can be extremely serious, up to, and including, criminal sanctions – not to mention termination of employment.

The simple line-in-the-sand with research funding, regardless of its source, is that it is intended for the purposes of conducting research and not, through nefarious means, as a form of personal *bonus* income for academic staff.

A basic tenet of research funding is that no academic can sign documents which would lead to direct benefits to himself/herself – or to willfully mislead a more senior staff member into approving such expenditure – without disclosing a vested, personal interest in the outcome.

Typical schemes which have been commonly established (and ultimately detected through audit) include:

- Establishment of artificial companies (by an academic grant-holder, spouse or colleague) for the purposes of invoicing a research grant, and thereby transferring funds to the grant-holder's personal finances.
- Use of funds for travel to conferences which are never actually attended in order to provide a vacation fund.
- Use of funds for personal entertainment or travel not directly related to the research project.
- Use of funds for non-research items which are of benefit to the grant holder (e.g., home entertainment system or other appliances).
- Use of research grant funds for the purchase of equipment/resources at an artificially inflated price, in exchange for back-hand payments from the supplier to the academic in charge of the grant expenditure.

Innovative as some naive academics may believe these schemes (and numerous others) to be, all can be readily revealed during financial audits, and the consequences can be dire. Unlike many other forms of misfeasance, these issues tend to be clear-cut and, because they are ultimately tied to the release/expenditure of university funds, are always fully documented – with physical or electronic signatures. This means that once the events are identified, prosecution against an individual staff member is highly probable.

13.9 Other Areas of Misfeasance

There are numerous other possible areas of misfeasance that a research supervisor may need to contend with during an academic career. These include:

- Cruelty/abuse of animals under laboratory care.
- Abuse/bullying/harassment of staff or students.
- Failure to comply with ethics procedures (e.g., publishing private information about survey respondents or medical information about experimental subjects).

These tend to be areas which are covered in significant detail by university procedures because they all relate to core university business, and breaches of faith in the conduct of that business. These issues have previously been covered in Chapters 3 and 9. All of these require a resolution approach similar to that which has been outlined in Table 13.1. In summary, this is:

- Act only on objective data, facts, statutory declarations or other written documents and correspondence.
- Do not make allegations based upon rumors, hearsay, verbal encounters or innuendo.
- As a first step, always work on the assumption of an innocent cause to the problems rather than misfeasance.
- Do not make allegations until all the evidence is in and, if possible, after the affected individual has been asked to respond to hard evidence.
- Maintain strict confidentiality and only communicate with those who are authorized to deal with privileged information regarding claims.
- Keep in mind that any publicly-aired allegations about misfeasance could lead to legal action against the person making the claims – particularly if they are slanderous or libelous.

14

The Long Term Supervisor/Researcher Relationship

14.1 Overview

Traditionally, the completion of a postgraduate research degree was considered to be the end of an apprenticeship that had been served, and the recipient was deemed to be worthy of an entry-level research or academic position within the university. The reality is that this is no longer the case.

The sheer volume of postgraduate research completions at a global level, relative to annually available academic positions, means that many people who complete their postgraduate research degrees will never work in the university system. Some will never work as researchers in industry, and some may never even work in a professional capacity.

The point here is that, where a postgraduate research degree once prepared a candidate for a particular type of career, the same is no longer true in the modern world. All research supervisors therefore need to think long and hard about the implications of this for their research students.

From a technical point of view, the research supervisor's role has effectively ended when the postgraduate research examination process has delivered a final verdict – hopefully a passing grade. Pragmatically, however, each research supervisor is also an academic staff member of the university, and therefore has an important, additional role as an institutional builder/advocate.

At this point it is opportune to reflect upon the Greek proverb (source anonymous) which says that,

> *"A society grows great when old men plant trees whose shade they know they shall never sit in."*

The same is true of universities. Each academic and researcher has the responsibility of planting the seeds for future generations of students, academics and researchers. Part of this involves accepting some responsibility for the future careers of those who pass through a university's portals. Part of this requires the building of long-term relationships with the student after graduation.

Needless to say, a research supervisor cannot guarantee the future career success of a research student any more than a university can guarantee the future success of a first-degree graduate. The key point here is that, in all cases, it is important that the graduates – be they first-degree or advanced research degree – leave confident that the university and its academics have done the best that they can to assist them in the transition towards a professional career.

For an undergraduate student making the transition from, say, an accounting degree to a career in accounting, the pathway is relatively straightforward, because the degree and the career are both well defined. The same is clearly not true of a research degree, where the future career pathways are many and varied, including:

- A postdoctoral position, teaching or research assistantship in the same university in which the research degree was completed.
- A postdoctoral position, teaching or research assistantship in a university other than the one in which the research degree was completed – possibly even in another country.
- A business/industry professional (non-research) position in a similar field to the one in which the research was conducted.
- A management career in business/industry/government.
- A commercial research and development career in business/industry.
- A start-up company, based upon ideas or technologies arising from the postgraduate research degree or other interests.

These aspects have already been discussed in Section 2.14 (Professional Career Foundation) and Section 3.9 (Professional Development). The primary issue to be addressed by the supervisor is that the postgraduate research degree does not, of itself, necessarily provide an expedited pathway into any of these career options. In each case, extracurricular efforts are required on the part of the research student, and support is required from the supervisor. Table 3.2 is reproduced here (as Table 14.1) to highlight the range of issues that need to be addressed.

The main point to take from Table 14.1 is that all of the possible career options and their implications need to be considered by the supervisor early in the research program – possibly even before the research student knows which direction he/she intends to take at the end of the program. Each potential career pathway necessarily involves some planning and preparatory work and,

Table 14.1 Example of possible development pathways and requirements

Career Aspirations	Key Factors	Timeframe
Academic Career	• Significant research outcomes • Publications/conference attendance • Citations • Academic service work – reviewing for journals, contributing to student programs • Networking with senior staff in current and prospective universities • Understanding of broader university research and education requirements and performance issues	3 Years
Professional Career in Broad Field of Research	• Understanding of research outcomes in the context of business/industry • Experience in/exposure to (e.g., internship) relevant companies • Understanding of company business models in fields of interest • Understanding of competitive advantages of PhD qualification relative to other graduates	1–2 Years
Management Career	• Formal management training/accreditation (e.g., MBA) • Basic business/industry experience (e.g., internship) in possible future employment organizations • High level communication ability • Networking with business-oriented colleagues	3 Years
Commercial Research and Development Career	• Demonstrable ability to deliver tangible research outcomes – on time • Ability to communicate R&D outcomes in an efficient, concise manner • Experience in commercial R&D environment (e.g., internship) • Possible Management training/accreditation (e.g., MBA)	2 Years
Start-up Company	• Understanding of basic entrepreneurship concepts and processes • Understanding of the business value proposition of the research outcomes • High level of communication ability • Networking with business angels, seed funding and venture capital organizations • Networking with enthusiastic colleagues to form possible teams	3 Years

by the time the research program has concluded, many of the opportunities for conducting this preparatory work will have already lapsed.

In this chapter, each of these possible pathways is examined from the perspective of how a supervisor can provide support to a research student in achieving such outcomes. At the end of the research program, the research student should then, at a minimum, have a profound appreciation for what the supervisor has done – potentially forming the basis for a long-term relationship.

In addition to these important *career-kick-starting* issues, there are also other basic things that a supervisor can do to help maintain an ongoing professional relationship with the research graduate, and these are also examined in this chapter.

The long-term benefits of a good relationship between the graduated student and the research supervisor should be clear – especially if the graduate becomes successful as an eminent researcher, a business owner in his/her own right, or a senior business/industry executive. The benefits can include:

- Donations to the university.
- Establishment of research collaborations.
- Funding of contract research and development at the university.

All these benefits can be substantially larger and more rewarding than any short-term benefits that a supervisor may gain during the course of the postgraduate program. All of the benefits that may accrue to the supervisor and the university have the same starting point – that is, doing everything possible to assist the research student in becoming successful in his/her chosen career pathway.

14.2 Academic Career

Traditionally, the logical extension to a postgraduate research degree has been a career in the university system. Unfortunately, the annual disparity between the number of people achieving postgraduate research degrees and the number of available tenured and non-tenured positions is so great that there is no guarantee that a graduate will ever enter academia.

A research supervisor's task is not necessarily to discourage a postgraduate student with the daunting statistics but, at the very least, to ensure that the student is aware that an academic career is not *fait accompli* after graduation. The objective is to get the point across that the competition for entry-level academic positions is enormous and, thereafter, the competition for tenured positions is also considerable. A research student therefore needs to understand that if he/she wishes to pursue a career in academia, then considerable preparatory work needs to be performed during the course of the postgraduate research program.

There are numerous tasks/achievements/milestones during the course of the postgraduate program that could tilt the statistics in the research student's favor. Some of these include:

- *Landmark research outcomes* – there is a great deal of serendipity involved in achieving benchmark research that becomes internationally acclaimed and cited – nevertheless, raising the sights of students over and above the mundane may inspire them to achieve significant outcomes. Quality always needs to take precedence over quantity in terms of research outputs.
- *Publications in respected, peer-reviewed journals* – each field has its own journals which are regarded as the benchmarks for acceptance of particular ideas – acceptance by peers in these journals is important for future career considerations in academia.
- *Citations* – highly cited work is well regarded in academia but, unfortunately, time constraints are such that most postgraduate research students will not have had their work exposed for a sufficiently long time to attract large numbers of citations.
- *Understanding the university environment and driving factors* – universities are large, complex organizations with diverse objectives and constraints – people who understand these are more likely to fit into the culture following the completion of their postgraduate degree. It is also important for the student to understand institutional research strengths and gaps in order to potentially fit into emerging academic positions.
- *Networking with a broad range of academics* – the research supervisor may not have the grants/funding to retain a research student as a post-doctoral scholar after graduation but other colleagues may – a research supervisor needs to ensure that his/her research students are networking with people who have the potential to provide academic career openings within the university or at other universities. Networking means more than just socializing – students have to be able to offer something of value and substance to those with whom they interact.
- *Experience in lecturing, laboratory supervision, tutoring, mentoring* – all basic academic functions, and ones in which experience may count when seeking a position.

Another consideration for the research supervisor is to try and ascertain the research student's strengths and professional preferences – and how these can best fit into the university environment. In particular, many research students may have a penchant for design, conduct and construction of experiments and experimental systems. These research students may prefer to have a university career in laboratory or technical support, rather than traditional academia or research. In large, modern universities, the career pathways for laboratory and technical staff can be both engaging and challenging – and provide an excellent alternative for high caliber graduates who prefer hands-on experimentation to broader research or academic roles.

Some research supervisors – either through naivety or a desire to hold on to a high caliber postgraduate student after graduation – often make grandiose promises about potential future employment in the same research group. Unless the supervisor has a well-established track record of competitive research grant income, or some other guarantees on the part of the university, it is best not to overstate the possibility of future employment in the same group, only to disappoint the student later. If one is intent on building up a long-term future relationship with the graduated student, failing to deliver at the first hurdle is not a good start.

A research supervisor, as part of his/her role as an educator, should be looking out for the best possible future career opportunities for his/her research student. This can involve introducing the student to a range of colleagues within the university or other various institutions – in the full knowledge that this may ultimately lead to losing the graduated student for potential postdoctoral positions.

All these issues require engagement and support from the research supervisor early in the research program, and the supervisor needs to ensure that meetings with a research student do not focus solely on the research *per se* but also upon the manner in which the student's research will manifest itself in future career options.

14.3 Professional Career in Broad Field of Research

Not all research students seek to become researchers when they graduate from an advanced degree. Some undertake the degree to pursue an area of special interest and others merely for the challenge of completing an academic program at the highest possible level. Once such a program is completed, a proportion of graduates will seek to move into a conventional professional career in the broader field of interest that has motivated their research. For example, someone who has undertaken a Doctoral degree in an area of bioscience may elect to move into the pharmaceutical industry – not as a researcher, but as a professional within the organization.

The obvious question then is, if students choose such a pathway, what is the role of the supervisor?

To begin with, it needs to be understood that those who undertake a postgraduate research degree, and then move into a conventional professional environment, are commencing their career outside the university at a considerable disadvantage. Postgraduate-qualified individuals have sacrificed several years of salary in order to achieve their qualification. Not only this, but their peers who moved directly into the professional market will likely have earned one or more promotions before the postgraduate qualified individuals even gain entry-level positions.

Postgraduate qualified individuals also incur opportunity costs as a result of being out of the general income-earning environment. For example, a professional with an income stream could purchase a property or stocks, and benefit from several years of capital growth, while a postgraduate research student does not have the means to undertake such investment.

If a research supervisor is seeking to develop a long-term professional relationship with the graduated student, then clearly sending that student out into the professional job market and having them accept a salary considerably lower than their already-promoted peers, who do not have postgraduate qualifications, is not a good start. A research supervisor therefore needs to take an active interest in the sorts of competitive advantages that a postgraduate qualification can contribute to the professional job market.

It is naive in the extreme for postgraduate research supervisors to delude themselves that their task is complete provided that their research students have undertaken quality research and have received their degrees. The reality in the modern world is that this simply isn't enough, because the postgraduate qualification, of itself, is insufficient to create the competitive advantage that the graduate will require to achieve high-caliber positions in business/industry.

A research supervisor therefore needs to become actively involved in the potential future career of his/her student, specifically:

- What sorts of companies hire students with advanced degrees?
- What are the competitive advantages (if any) for people with advanced degrees in those companies?
- How can the research supervisor maximize his/her student's competitive advantages?

- What additional training or extra-curricula activities will the student need to undertake in order to become a more convincing value-proposition for potential employers?
- Who are the recruitment decision-makers in organizations that are potential employers?
- How can the supervisor get the research student to network with potential employers and decision-makers, or else bring those decision-makers to the university to showcase his/her student's work?

These are all important questions that the research supervisor needs to address, once he/she understands the research student's predisposition to a particular future career. The objective is to make the research student as valuable as he/she can be so that he/she can have the best opportunity for a professional career in the field of interest.

Many research supervisors have little or no exposure to business or industry themselves and so, unsurprisingly, they are hesitant to get involved in something on behalf of their students, which they themselves find difficult.

There are two factors working in the favor of research supervisors in this regard. The first is the credibility/brand-value of the university itself in helping the supervisor forge an introduction. The second is the machinery that is already in place at the university for recruitment of undergraduate students.

Cold-calling external organizations, to forge relationships, is less onerous for a university staff-member than for a student or member of the public. Universities are seen to be organizations of public good, and many business and industry people genuinely welcome the opportunity of supporting the educational process. Some organizations are also keen to recruit high caliber graduates, and so any mechanism that the research supervisor can create to build bridges, and demonstrate the abilities of his/her postgraduate research students to business/industry is a value-adding activity.

Each university has some form of outreach program or graduate recruitment program that it conducts at a broad level with business, government and industry. The connections and introductions that a research supervisor will require in order to assist his/her student in networking are therefore generally already available within the system. The research supervisor should make use of these to forge programs that bring relevant potential employers closer to postgraduate students who see their futures in business or industry.

Finally, most large business/industry organizations have internal recruitment coordinators and departments specifically set up to interact with universities and their students. A research supervisor can also make use of these to help the student make the necessary linkages.

Importantly, there is more to supporting a research student achieve his/her career objectives than making a simple introduction or handing out a business card. The research supervisor needs to work with the student in understanding what potential employers want, and what factors/triggers can lead to:

- Greater desirability.
- Higher entry salaries.
- Broader position scope.

Sometimes, the supervisor's maturity and life experience can, of itself, be a great benefit to helping a research student bridge the gap between his/her research and a potential career pathway with a particular commercial or government entity.

Clearly, these are all issues that need to be worked upon from the early days of the postgraduate research program, through to the conclusion and ultimate recruitment.

14.4 Management Career

A postgraduate research degree may seem incongruous with a future management position in business or industry. In the classic XY-graph of *breadth vs depth* that is sometimes applied in the commercial environment, those who have a great depth and limited breadth are deemed technocrats – and those who have great breadth but limited depth are deemed managers. The two positions appear to be incompatible.

The *breadth vs depth* model applies more appropriately to large, traditional companies than it does to lean, modern organizations with limited resources and a limited operating life. In these leaner organizations, there are potentially management roles for those with a large depth of knowledge – provided that they are prepared to put in the work to understand the semantics and semaphores of business and industry.

There is only so much that a research supervisor can do in isolation in order to prepare a research student for a management pathway. Clearly, additional training, beyond the postgraduate research program, is required. This could take on the form of a traditional Master of Business Administration (MBA) or other forms of training and accreditation in business and entrepreneurship.

In coming to an understanding of where the research student would like his/her career to go, beyond the research degree, the supervisor needs to consider how other supporting training can be provided – either within the university or at another appropriate institution. This will also involve additional workload for the research student, and the supervisor needs to understand the impact that this may have upon the research program.

In addition to any basic accreditation in the business field, there are also mindset changes that may need to take place. In particular, the objective of postgraduate research is to foster a thought process where attention is given to the analysis of every small detail – in order to eliminate uncertainty. In the business/industry environment, there is seldom sufficient time to eliminate uncertainty in the decision-making process. Decisions often need to be made at a visceral level by balancing risks and rewards, in an environment where the data is manifest in shades of gray rather than black and white.

Is it practical for a research supervisor to inculcate such a mindset shift on the part of a research student? The answer to this question clearly depends upon the research student. Some research students are able to quickly and naturally adapt from one form of thinking to another. Others are not.

It may be that those who are steadfastly wed to the consideration of detailed analysis of data and information in order to make decisions will never become senior managers – but they may be the material of great researchers. It may also be that those who can shift from detailed analysis to broad decision-making in an environment with shades of gray will never become great researchers, but they may become exceptionally good technology managers.

At some point, in coming to know the research student, as one does during the course of a supervision, the research supervisor needs to determine which of the personality traits his/her student has, and how that will affect future career aspirations.

Other factors will also influence potential pathways to management. For example,

- Is the research student gregarious, shy or a loner?
- Does the research student naturally lead when placed into a team environment?
- Can the research student bring together a disparate band of individuals and make a team which is greater than the sum of its parts?

These are the sorts of questions that need to be answered in order for the supervisor to get an insight into whether his/her student is suited to a management pathway.

A good way to determine the particular strengths and weaknesses of a student is to provide opportunities for team-based activities – whether these be formal academic activities, informal academic exercises or informal extra-curricula activities. It may be that, in providing such opportunities, the research student will be able to come to his/her own conclusions about suitability for a management pathway, and the added commitment and potential training that will be required in order to fulfill such ambitions.

14.5 Commercial Research and Development Career

A career in commercial research and development (R&D) may appear to be as logical an extension to a postgraduate research program as an academic research career. However, there are significant differences and, because competition for commercial R&D positions is intense, research supervisors may have an important role to play in maximizing the chances of their research students achieving success in such positions.

In the middle of the 20th Century, many large, multinational corporations maintained *blue-sky* R&D facilities (e.g., IBM and Xerox), wherein groups of high-caliber graduates and researchers had the freedom to tackle challenging problems that might only have been peripherally related to a corporation's core business activities. Some of these activities led to commercially significant products, and some created collateral technologies that were eventually incorporated into other mainstream products. In the modern world, however, the cost of maintaining such facilities is often prohibitive, particularly because of tight profit margins, limitations on cash flow, and so on. Many modern commercial organizations have therefore developed a portfolio approach to the management of commercial R&D.

A portfolio approach to commercial R&D means that the manner in which research activities are tackled is segmented according to risk (or probability of success). For example:

- *Low-risk activities*, which are likely to yield a commercial outcome within a short timeframe, are generally conducted in-house with full ownership of intellectual property (IP) retained by the company.
- *Medium-risk activities*, which have a reasonable chance of a commercial outcome, but which may involve too large a risk in terms of investment, may be performed in partnership with other commercial organizations or universities.
- *High-risk activities*, which have only a minute chance of success, and for which commercialization mechanisms *are unclear*, are farmed out to universities and other government R&D facilities through research grants, and only a portion of the total IP is retained by the company in the event of success.

- *High-risk activities*, which have only a minute chance of success and for which commercialization pathways *are clearly identifiable*, may be fostered through venture capital or seed funding of innovative researchers – with the prospect of buying out the resulting company if it proves to be successful – thereby effectively outsourcing high-risk R&D.

A good example of this portfolio approach might be in the pharmaceutical industry, where research into minor variations to mainstream products is performed in-house, while new pharmaceutical products, with a high degree of risk in terms of development and certification, are left to small biotechnology companies. The small start-ups wear the risk through venture capital investments (sometimes through mainstream pharmaceutical companies) and possibly other supporting funding from government.

The key point to take from all this is that a postgraduate research student cannot simply expect to pursue the same research pathway in modern commercial research as in the university environment. There are significant pressures to deliver tangible results – on time and on budget. Moreover, in commercial research, the emphasis is on team-based outcomes, where individuals are generally not permitted to lay claim to significant portions of IP – in order to prevent individuals from starting up competitor organizations.

In order to tackle these cultural differences, the research student needs to understand basic principles of:

- Entrepreneurship.
- Management of R&D in the corporate environment.
- Basic business management.
- Team-based research.

It may not be necessary (or feasible) for an individual to possess additional, formal qualifications in these areas in order to get a foot in the door of commercial R&D, but extensive reading on the subject could be a particularly useful competitive advantage.

From the supervisor's perspective, there are various things which can be done to assist the student in making a transition to commercial R&D. Bringing industry people into the university to see research first-hand is a good starting point. Exposing the research student to staff in other relevant faculties is also useful – especially if the university has faculties which have a focus on entrepreneurship; venture capital, and so on.

At a more basic level, in terms of research supervision, it may be worthwhile to provide a much greater focus on research planning and management on the part of the student – particularly in areas such as project timelines, critical path management, and so on. It may also be worthwhile for the student to engage with potential employer organizations and, perhaps, work as an intern in order to gain experience in the commercial environment.

A good research supervisor should have an awareness of the companies that recruit researchers in his/her field – and should work to establish ongoing relationships with those companies. This can foster a greater synergy between commercial research departments and the work undertaken at the university, and also allow commercial organizations to develop some trust in the research staff and students at the university. With a good relationship intact, the research supervisor may become a first port of call for companies looking to recruit postgraduate research qualified staff.

14.6 Start-Up Company

Many research students in the modern world, who may be frustrated by limited, exciting prospects in traditional research environments, see the possibility of establishing a start-up company, based upon their university research, as a means of retaining a foothold in the area of interest, as well as generating an income.

The probability of start-up companies, initiated by postgraduate research students, becoming long-term successful is very small, but even a failed enterprise can provide an entrepreneurial graduate with invaluable experience – to either start up another company or work in other commercial areas.

The notion of a start-up company being a pathway to riches is a rule borne out in the exception. Nevertheless, the allure of financial success is often a powerful incentive for graduates to work tirelessly in setting up a new enterprise.

For the research supervisor, providing support to a research student in starting up his/her own enterprise can have numerous benefits for a long-term professional relationship after the conclusion of the postgraduate program.

As a starting point, if a research student's commercial venture becomes successful, and the academic supervisor is seen as a person who was instrumental in supporting that success, then numerous benefits can flow back to the university and the supervisor – including contract R&D, research project funding, donations and support for university activities. Even if a venture is unsuccessful, any reasonable graduate should still recognize the efforts and contributions that his/her supervisor has made in trying to create that success, and various collateral benefits may flow from this.

A research supervisor whose student has expressed an interest in using his/her postgraduate work as the basis of a company can provide numerous supporting functions to that student. First and foremost, on behalf of the student, the supervisor needs to formally determine the status of IP arising from a research student's project. If a postgraduate student has been funded through a collaborative research project with industry, or if the student has received a scholarship from industry, then formal agreements may be in place in relation to the disbursement of IP arising from the project. In some universities, it may also be the case that the university itself owns the IP resulting from a postgraduate research project. If this is the case, then the student has no *prima facie* basis on which to start a company. It may be possible, however, for a research supervisor to work with relevant partners, and the university's nominated representatives, to negotiate a mechanism by which a start-up company may be formed – perhaps, this might be through the allocation of equity in the new organization.

Assuming that a start-up company can be formed, it needs to have a more systematic driving force than simply a research student with boundless enthusiasm. To begin with, in order to make meaningful headway in the realm of start-up companies, a postgraduate student will require:

- Training/education in the field of entrepreneurship.
- Advice/support from business/venture capital and legal professionals.
- Seed funding and venture capital funding.
- Office accommodation/space – incubation area.
- Basic infrastructure – information technology, payroll, accounting software, etc.

Clearly, most supervisors are not equipped to themselves provide this type of support. However, it is not unreasonable to expect that supervisors should make an effort in informing themselves on the best means by which such resources can be obtained – particularly if there are programs available within the university to provide them.

Some universities are better equipped than others to support the start-up activities of research students. For example, an international leader in university-based start-up activities, Stanford University, provides support through its StartX Accelerator Program *(Startx.com, 2015)*, which addresses many of the above issues on behalf of research students, and without placing an unnecessary equity burden on the newly formed start-up.

Needless to say, not all universities can provide the Stanford level of support to their research students, but each university will have some elements available in-house, which a research supervisor may be able to tap into on behalf of his/her students. For example, a research supervisor may be able to negotiate incubation space for a start-up, including use of the university's IT infrastructure. Another possibility may involve tapping into the goodwill of academic colleagues in accounting/business/economics fields in order to provide a team of mentors for the research student, with expertise broader than just the core knowledge itself.

It would be all too easy for a research supervisor to shrug his/her shoulders and complain that these sorts of activities are not his/her responsibility. However, this is ignoring the realities of modern supervisory practice – regardless of whether such additions are enshrined in university procedures.

In addition to these supporting activities, a research supervisor may also wish to take on an additional mentoring role in the context of trying to evaluate the research student's personality and work practices against those required for an entrepreneurial start-up company. If a supervisor feels that his/her research student simply doesn't possess a team-building capacity, or is not naturally gregarious, then there is some responsibility to support the student by identifying deficiencies/weaknesses and seeing how they might be worked upon in order to become strengths.

Finally, there is the notion of a supervisor investing his/her own personal money into a start-up company to support the research student. Some highly successful technology companies started on just such a basis. As a general rule, however, this might be overstepping the boundary of an educator. Additionally, if a research supervisor is not an informed and experienced investor, it may lead to the ultimate loss of personal savings. Over and above these issues is the possibility that, by investing in a research student's company, the supervisor may establish a conflict of interest by holding a personal equity in an organization that may also attract other funding or resources from the university itself.

According to Fortune Magazine *(Griffith, 2014)*, nine out of ten start-up companies fail. The key reason for failure is a product for which there is no real market,

> "...CB Insights recently parsed 101 post-mortem essays by startup founders to pinpoint the reasons they believe their company failed. On Thursday the company crunched the numbers to reveal that the number-one reason for failure, cited by 42% of polled startups, is the lack of a market need for their product."

Entrepreneurship is largely founded on the principle of having the ability to predict what the market will want before that market knows what it wants. This was well encapsulated in the words of automotive industrialist and entrepreneur Henry Ford, who said,

> *"If I had asked people what they had wanted, they would have said faster horses."*

In the context of university postgraduate research, however, it may be that the technological/social boundaries of what is being created by research students are simply too far ahead of the market to facilitate short-term commercialization. In other words, within the normal operating window of a start-up company, many university research outcomes may not have a market for some years – or even decades.

These aren't decisions that a research supervisor can make on behalf of a research student, beyond providing impartial, sage advice on the perils of investing heavily in a venture which is inherently risky. The research supervisor's role is neither to dampen enthusiasm nor overstate the benefits of the start-up option. What is called for is a reasoned and informed presentation of the evidence in relation to start-ups and entrepreneurial traits that lead to success. Beyond that, there is a need for the supervisor to respect the research student's ultimate decision and provide as much support as possible to maximize the chances of success.

14.7 Other Basic Relationship Building Issues

Thus far, this chapter has documented only the relationship building issues that are associated with a research student's career choices. There are, however, more basic means by which a research supervisor can establish and maintain a long-term professional relationship with a research student. The obvious ones include:

- Involvement and participation in university alumni organization functions.
- Organization of formal and informal functions for graduated research students to attend on a regular basis.
- Inviting graduates to lunch or dinner to maintain contact with them and follow their career choices.

14.7 Other Basic Relationship Building Issues

These should already be apparent to anyone who has reached the stage of becoming a research supervisor. Nevertheless, it is important to understand that these sorts of functions can only work if the graduate student:

- Has a genuine and profound respect for his/her supervisor – specifically, as it relates to integrity, knowledge and understanding.
- Believes that the supervisor has the ability to provide ongoing value (i.e., knowledge or advice) to his/her career.
- Believes that, at all times during the supervisory process, the research supervisor put the student's interests ahead of his/her own personal and career interests.

Without these underlying qualities, many of the mechanisms that are in place within the university system to maintain ongoing relationships with students may appear shallow and insincere. Simply put, the research supervisor can neither buy friendship nor respect – both of these need to be hard earned over the years spent with the research student during the postgraduate research program.

The long-term relationship between the research student and supervisor cannot be built at the end of the research program – it needs to be built from the very beginning through mutual respect and trust.

15

Questions Supervisors Should Ask Themselves

15.1 Overview

It should be apparent to those who have read the preceding chapters of this book that supervising a research candidate is far more complex than just providing field-specific technical advice in an area of specialization. Those academics who choose to undertake research supervision carry considerable responsibilities – ethical, moral and legal.

Research supervisors have the luxury of working in an environment which has significant flexibilities and graces that are not normally accorded to professionals in the commercial environment. With flexibility, however, comes added responsibility. This is particularly important in the context of research supervision, because the process involves a supervisor exerting significant control over the work, wellbeing and future career of a junior professional.

It is therefore worthwhile to conclude this book by posing a number of questions for potential research supervisors – which each individual can only answer for themselves in the light of what they have read herein.

15.2 What Is My Motivation for Research Supervision?

The motivation for postgraduate research supervision is particularly important. Some academics take on research students solely for the purpose of advancing a specific aspect of knowledge – in other words, as a tool to be used in the process of discovery. Other academics take on students in order to teach them the profession/craft of research in a particular field. Different motivations can lead to vastly different outcomes in the context of the research student.

A research student whose supervisor only wants research outcomes may not learn the discipline and process of research. In effect, that student may be acting merely as a research assistant or laboratory technician, and the assumption is that, in so doing, he/she will automatically pick up the discipline itself. This assumption may be ill-founded.

A research student whose supervisor only wishes to impart the discipline of research may become a highly competent research practitioner but, perhaps, one where adherence to process overshadows the challenging of established paradigms and artificial knowledge boundaries.

Research supervisors need to think long and hard about what has motivated them to supervise a student, and how the focus of the supervision can be to create both a competent research practitioner and one who can also be a free-thinking paradigm shifter.

15.3 How Will My Personal Ambition Impact on Supervision?

Supervisors need to understand what drives them in the academic/research context. Specifically, are they driven by:

- An ambition for academic or personal career?
- A desire for fame through international renown – by making major research breakthroughs?
- A thirst for knowledge discovery?
- The excitement of a particular field?
- A desire to educate and shape minds?

The answer to this question is particularly relevant, and it is critical that each individual addresses it honestly in his/her own mind before starting a research supervision.

There is nothing wrong with personal ambition, or a thirst for knowledge, or enthusiasm for a field. The problem is that when one is charged with the responsibility of educating another individual – in a highly specialized field – it is important that each decision is made with the best interests of that individual in mind. In other words, the research student must rank above personal career ambition, discovery and even thirst for knowledge.

15.4 How Do I View My Role in the Research World?

Some academics and researchers get enjoyment from the research journey – that is, exploring each pathway and reporting factually the promise that each pathway holds for future researchers. Other academics and researchers are primarily interested in the end-goal – that is, curing a disease, creating a more efficient form of energy generation, and so on.

Some academics and researchers are pragmatists – and understand that if they are not the ones serendipitous enough to make a discovery, then others eventually will anyway. Other academics are messianic, and believe that it is their role in life to achieve a particular research breakthrough.

How a supervisor views his/her role in research is important in the context of research supervision. A messianic person will exert a high level of pressure on a research student to achieve particular goals, with little interest in how they are achieved or how the pressure affects the student. Those that get enjoyment from the research journey will create a different kind of graduate altogether – one that may have a genuine interest in exploration.

The reality is that the world needs all these different kinds of researchers. Some to lay the groundwork, and others to force knowledge to cross established – but artificial – boundaries. However, each supervisor needs to understand where his/her preferences are, and how they will impact upon the research student. It is important also that they explain their preferences to the student – preferably before a research program commences – so that a student can decide whether or not they can live with the traits of the supervisor.

15.5 What Is My View on Research Supervision?

Some supervisors like to adopt a laissez-faire approach to supervision, wherein the research student is master of his/her own destiny. The research student provides the drive and the supervisor provides only gentle background advice when asked. Some supervisors like to adopt a master-apprentice model, wherein the supervisor is the major player in the research, and the student is a research assistant who learns by watching the master at work.

There are advantages and disadvantages to both types of supervision but it is important a supervisor understands his/her own preferences and conveys them to the student so there is shared clarity in purpose.

15.6 What Sort of Professional Traits Do I Have?

Some academics/researchers are introverted and like to work in isolation – others like to work in a team environment and share responsibilities and information. Some academics/researchers are gregarious while others are timid and shy. Some academics/researchers are institutional builders while others are personal career builders. Some academics/researchers are bridge-builders (*out-reachers*), who like to project their work outwards to business and industry – others like to maintain strict controls in-house. Some researchers are pedantic (*details*) people and others are paradigm-shift (*big-picture*) people.

From a supervisory perspective, each academic/researcher needs to know what his/her professional traits and strengths/weaknesses are, and how these may impinge upon the research student – both in terms of the research program and future career.

If a research student is intent on pursuing a career in pure (basic) research, then they may appreciate a supervisor who is focused strictly on individual excellence and academic matters. If a research student is intent on creating a start-up company at the end of the research program, then he/she may appreciate an entrepreneurial supervisor who is an established bridge-builder with external, commercial organizations. If there is a mismatch between research student aspirations and supervisor traits, then problems and conflicts may ensue.

15.7 Am I Currently in a Good Position to Supervise a Student?

Research supervision is not a short-term activity. It requires several years of genuine commitment. If something happens to a supervisor, then it is difficult to substitute another academic with differing ideas on how a research program should be conducted. This can be very unfair and confusing to a research student. A potential supervisor therefore needs to undertake a situation report on their own career before agreeing to supervise a candidate.

Consider that an academic may have a career which is itself unstable. The current university position may be untenured, and a contract may expire before a research student has completed his/her degree. Perhaps there is a tempting job offer elsewhere in the university world. How would it impact upon the research student if the supervisor was to leave his/her position? Unlike most other forms of professional relationship – say, employer-employee – one cannot simply swap out the parts and attain parity. A potential supervisor needs to consider this.

If some degree of instability is inevitable or unavoidable, then a potential supervisor needs to have a contingency plan for the student. This is the minimum that one would expect of a professional supervisor, and it also behooves that supervisor to advise his/her students of possible instabilities at the earliest possible occasion in the program.

15.8 The Most Important Question

In order to make a postgraduate research program successful in the broadest context – that is, a high quality research outcome and a highly employable or successful graduate, there is one basic question, above all others, that supervisors need to address. Specifically,

> *"Can I put to one side all my personal and research ambitions, and career aspirations, and commit myself to always acting in the best interests of the research student as a first priority?"*

If, after significant introspection, the answer to this question is *yes*, then there is a solid foundation for supervising the research and for resolving the many problems that will inevitably arise.

If the answer to the question is *no*, then one has to accept that some of the problems which will arise during the conduct of the research program will become insoluble and the likelihood of a successful program conclusion will be reduced.

George Washington once summarized profoundly the need for integrity in life and work when he wrote,

> *"Labor to keep alive in your breast that little spark of celestial fire called conscience."*

Academics may benefit from this credo and consider it a sound basis for decision-making as they tackle the inherent challenges of research supervision.

Appendix

Bibliography/References

[1] Bush, V. (1945). *Science, the Endless Frontier*. [online] Nsf.gov. Available at: http://www.nsf.gov/od/lpa/nsf50/vbush1945.htm#ch3.6 [Accessed 5 Sep. 2015].

[2] Merriam-webster.com, (2015). *null hypothesis*. [online] Available at: http://www.merriam-webster.com/dictionary/null%20hypothesis [Accessed 27 Sep. 2015].

[3] Stokes, D. (1997). *Pasteur's Quadrant*. Washington, D.C.: Brookings Institution Press.

[4] Quoteinvestigator.com, (2012). *If I Had More Time, I Would Have Written a Shorter Letter | Quote Investigator*. [online] Available at: http://quoteinvestigator.com/2012/04/28/shorter-letter/ [Accessed 7 Sep. 2015].

[5] Shanghairanking.com, (2015). *Academic Ranking of World Universities – 2015 | World University Ranking – 2015 | Top 500 universities | Shanghai Ranking – 2015*. [online] Available at: http://www.shanghairanking.com/ARWU2015.html [Accessed 5 Sep. 2015].

[6] Ets.org, (2015). *TOEFL iBT: Understand Scores*. [online] Available at: http://www.ets.org/toefl/ibt/scores/understand/ [Accessed 5 Sep. 2015].

[7] Ielts.org, (2015). *IELTS | Institutions – IELTS band scores*. [online] Available at: http://www.ielts.org/institutions/test_format_and_results/ielts_band_scores.aspx [Accessed 5 Sep. 2015].

[8] DeSantis, N. (2012). *You Can Summarize Your Thesis in a Tweet, but Should You? – Wired Campus – Blogs – The Chronicle of Higher Education*. [online] Chronicle.com. Available at: http://chronicle.com/blogs/wiredcampus/you-can-summarize-your-thesis-in-a-tweet-but-should-you/34962 [Accessed 14 Mar. 2016].

[9] A Blog Around The Clock. (2016). *Using Twitter to learn economy of words – try to summarize your research paper in 140 characters or less!*. [online] Available at: http://scienceblogs.com/clock/2010/02/28/using-twitter-to-learn-economy/ [Accessed 14 Mar. 2016].

[10] Threeminutethesis.org, (2015). *About*. [online] Available at: http://threeminutethesis.org/about-3mt [Accessed 7 Sep. 2015].

[11] AllTrials, (2015). *What does all trials registered and reported mean?* [online] Available at: http://www.alltrials.net/find-out-more/all-trials/ [Accessed 8 Sep. 2015].

[12] Americanrhetoric.com, (2015). *American Rhetoric: Russell Conwell – "Acres of Diamonds"*. [online] Available at: http://www.americanrhetoric.com/speeches/rconwellacresofdiamonds.htm [Accessed 10 Sep. 2015].

[13] WHO.int, (2015). *WHO | Occupational health*. [online] Available at: http://www.who.int/topics/occupational_health/en/ [Accessed 11 Sep. 2015].

[14] WHO.int, (2015). *WHO | Occupational health*. [online] Available at: http://www.who.int/topics/occupational_health/en/ [Accessed 11 Sep. 2015].

[15] Counselling.cam.ac.uk, (2015). *Reducing the risk of student suicide – University Counselling Service*. [online] Available at: http://www.counselling.cam.ac.uk/staffcouns/leaflets/suiciderisk [Accessed 12 Sep. 2015].

[16] Citiprogram.org, (2015). *CITI – Collaborative Institutional Training Initiative*. [online] Available at: https://www.citiprogram.org/index.cfm?pageID=91 [Accessed 12 Sep. 2015].

[17] Citiprogram.org, (2015). *CITI – Collaborative Institutional Training Initiative*. [online] Available at: https://www.citiprogram.org/index.cfm?pageID=88 [Accessed 12 Sep. 2015].

[18] Definitions.uslegal.com, (2015). *Harassment Law & Legal Definition*. [online] Available at: http://definitions.uslegal.com/h/harassment/ [Accessed 13 Sep. 2015].

[19] Definitions.uslegal.com, (2015). *Bullying Law & Legal Definition*. [online] Available at: http://definitions.uslegal.com/h/bullying/ [Accessed 13 Sep. 2015].

[20] Thomas, K. and Kilmann, R. (1974). *Thomas-Kilmann Conflict Mode Instrument*. Consulting Psychologists Press.

[21] Kilmanndiagnostics.com, (2015). *An Overview of the TKI | Kilmann Diagnostics*. [online] Available at: http://www.kilmanndiagnostics.com/overview-thomas-kilmann-conflict-mode-instrument-tki [Accessed 20 Sep. 2015].

[22] Webometrics.info, (2015). *About Us | Ranking Web of Universities*. [online] Available at: http://webometrics.info/en/About_Us [Accessed 24 Sep. 2015].

[23] Shanghairanking.com, (2015). *Ranking Methodology of Academic Ranking of World Universities – 2015*. [online] Available at: http://www.shanghairanking.com/ARWU-Methodology-2015.html [Accessed 24 Sep. 2015].
[24] Kavliprize.org. (2016). *About The Prize*. [online] Available at: http://www.kavliprize.org/about [Accessed 15 Mar. 2016].
[25] Breakthroughprize.org. (2016). *Breakthrough Prize*. [online] Available at: https://breakthroughprize.org/ [Accessed 15 Mar. 2016].
[26] Thomson Reuters (2015). *Web of Science – IP & Science – Thomson Reuters*. [online] Wokinfo.com. Available at: http://wokinfo.com/ [Accessed 28 Sep. 2015].
[27] Nih.gov, (2015). *National Institutes of Health (NIH)*. [online] Available at: http://www.nih.gov/ [Accessed 28 Sep. 2015].
[28] NSF.gov, (2015). *National Science Foundation*. [online] Available at: http://www.nsf.gov/ [Accessed 28 Sep. 2015].
[29] Eisenhower.archives.gov, (2015). *Eisenhower Presidential Library*. [online] Available at: http://www.eisenhower.archives.gov/all_about_ike/quotes.html [Accessed 5 Oct. 2015].
[30] Johnson, S. (1765). *Mr. Johnson's Preface to His Edition of Shakespeare's Plays*. London: Printed for J. and R. Tonson, H. Woodfall, J. Rivington et al., p.lvii.
[31] Robson, J. (1984). *John Stuart Mill, The Collected Works of John Stuart Mill, Volume XXI – Essays on Equality, Law, and Education*. London: Routledge and Keegan.
[32] Thinkchecksubmit.org, (2015). *thinkchecksubmit*. [online] Available at: http://thinkchecksubmit.org/ [Accessed 6 Nov. 2015].
[33] Doaj.org, (2015). *Directory of Open Access Journals*. [online] Available at: https://doaj.org/ [Accessed 6 Nov. 2015].
[34] Alpsp.org, (2015). *ALPSP Home Page*. [online] Available at: http://www.alpsp.org/Ebusiness/Home.aspx [Accessed 6 Nov. 2015].
[35] Arxiv.org. (2016). *arXiv.org e-Print archive*. [online] Available at: http://arxiv.org/ [Accessed 24 Mar. 2016].
[36] Biorxiv.org. (2016). *bioRxiv.org – the preprint server for Biology*. [online] Available at: http://biorxiv.org/ [Accessed 24 Mar. 2016].
[37] Pubpeer.com (2016). *PubPeer – Search publications and join the conversation*. [online] Available at: https://pubpeer.com/ [Accessed 25 Mar. 2016].

[38] Sparc.arl.org, (2015). *SPARC*. [online] Available at: http://www.sparc.arl.org/ [Accessed 6 Nov. 2015].
[39] Startx.com, (2015). *StartX*. [online] Available at: http://startx.com/accelerator [Accessed 20 Dec. 2015].
[40] Griffith, E. (2014). *Startups are failing because they make products no one wants*. [online] Fortune. Available at: http://fortune.com/2014/09/25/why-startups-fail-according-to-their-founders/ [Accessed 20 Dec. 2015].

Index

1:10:100 Ratio (for research commercialization) 66

A
Academic Ranking of World Universities (ARWU) 97
Acres of Diamonds 53
AllTrials 43
arXiv (Cornell University) 218
Association of Learned and Professional Society Publishers 217

B
bioRxiv (Cold Spring Harbor Laboratory) 218
Blue-sky R&D 295
Bohr (Bohr Quadrant) 24
Breakthrough Prize 100
Bush, Vannevar 3

C
Candidature (research) 6, 124, 125, 169
Central research theme 28, 30, 229, 248
Closed hypothesis 22, 42
Cold Spring Harbor Laboratory 218
Collaborative research program 66, 117, 118, 282
Committees 125, 156, 169, 272
Consejo Superior de Investigaciones Científicas (CSIC) 96, 97
Conwell (The Reverend Russell) 53
Cornell University 218
Cost center(s) 104, 115, 181, 189

D
Discrimination, Harassment and Bullying 56, 69, 132, 162

Directory of Open Access Journals 217
Dissertation 7, 47, 132, 239
Doctorate 4, 80, 125, 256

E
Edison (Edison Quadrant) 24, 25
Einstein, Albert 4, 229
Eisenhower, Dwight 145
English language support units 237
Ethics (research) 67, 72, 127, 283
Examination process 76, 149, 213, 285

F
Feedback (Timely Feedback) 6, 40, 141, 219
Fee for publication (journal) 219
Fields Medal 52, 96, 99, 109
Ford, Henry 300
Fortune Magazine 299

H
Harvard University 68
Health (and Safety) 6, 131, 195, 199

I
IBM 295
Induction programs 58, 59, 60, 132
Industry-based Doctorates 4
Institutional funding 102, 103
Insurance (liability/injury) 181
International English Language Testing System (IELTS) 34

J
Johnson, Samuel 155
Journals (ranking) 21, 99, 214, 280

K
Kavli Prize 100
Kissinger, Henry 155

L
Lao Tzu 85
Leiden (university rankings) 97
Literature review 238, 241, 243, 245

M
Methodology 10, 47, 124, 233
Metrics (research) 52, 97, 109, 206
Mill, John Stuart 161
Misconduct 11, 167, 193
Misfeasance 162, 166, 265, 283

N
National Institutes of Health (NIH) 104, 275
National Science Foundation (NSF) 104
Nature (journal) 3, 57, 109, 202
Nobel Prize 52, 99, 100, 109
Null hypothesis 20, 22, 149

O
Open hypothesis 19, 22, 42, 149

P
Pascal, Blaise 28
Pasteur's Quadrant Diagram 24
Project planning 146, 147
Psychological welfare 60
PubPeer – research article checking mechanism 275

Q
Qualifying Examinations 4

R
Research examiners 25, 56, 76
Rote learning 16, 123, 138, 279

S
Scholarly Publishing and Academic Resources Coalition 217
Scholarship (funding) 10, 116, 118, 120
School(s) of Thought 35, 77, 223, 242
Seminal research papers 239
Start-up (company) 75, 129, 286, 297
Shanghai Jiao Tong – See Academic Ranking of World Universities 96, 97, 98
Socratic Method/Approach 39, 276
StartX Accelerator Program (Stanford University) 298
Strategic, Tactical and Operational (management levels) 108
Stress 56, 59, 137, 168
Suicide (Cambridge University Counseling Service possible causes) 62, 63
Swift, Jonathon 39

T
Tao te Ching 35
Teaching hospital 60, 173, 177, 183
Test of English as a Foreign Language (TOEFL) 34, 35
Times Higher Education (THE) University Rankings 97
Thesis 47, 221, 232, 261
Think. Check. Submit. Consortium 217
Thomson Reuters 98
Timeline of discovery 239, 241, 242, 246
Tutu, Desmond 158
Travel Costs 115, 116

U
USLegal 70, 71

W
Web of Science 98
Webometrics 96, 97

X
Xerox 295

About the Author

Dario Toncich was born in Melbourne Australia in 1960 and graduated, with honors in Electrical Engineering, from the University of Melbourne in 1983. His industry based research led to a Master of Engineering (by Research) from SIT and a PhD from SUT. Since that time, he has held industrial, academic and research leadership positions, and has been involved in numerous research and development programs. In 1988, Dr. Toncich was appointed as the foundation research manager for the (then) recently established Key Centre for Computer Integrated Manufacture (CIM). In 1996, when the Centre was expanded into an industrial research institute (IRIS), he went on to become the research leader in automation and control; subsequently headed the Institute's postgraduate research and coursework programs, and contributed towards its strategic planning activities.

Dr. Toncich has authored and coauthored a number of research papers in his field. He has also contributed towards several federal reviews of national research policy and expenditure. His views on industry research and development have been cited in government reviews and editorialized in Business Review Weekly. Dr. Toncich has supervised more than 20 Doctoral and Master's (by research) candidates to successful completion, and has authored five text books.

Dario Toncich's field of research and Doctoral supervision has been in the areas of automation and control, with emphasis on feedback signals and signature analysis for both industrial and biomedical devices. Dr. Toncich's views and papers on higher education and research practice have been cited by governments and universities and appeared regularly in Campus Review. In 2007 he was appointed as the Project Manager for the University-wide research degrees admissions project at Monash University and in 2010 he worked as the Research Performance Analyst for the Office of the Senior

Deputy Vice Chancellor and Deputy Vice Chancellor Research at Monash University, in the Research Strategy Unit.

Dario Toncich regularly provides problem-solving advice to postgraduate research supervisors and students who are experiencing difficulties, and also acts as an advisor on university research performance assessment and benchmarking with consulting firms.